Lecture Notes in Computer Science 606

Edited by G. Goos and J. Hartmanis

Advisory Board: W. Brauer D. Gries J. Stoer

Series Editors

Gerhard Goos
Universität Karlsruhe
Postfach 69 80
Vincenz-Priessnitz-Straße 1
W-7500 Karlsruhe, FRG

Juris Hartmanis
Department of Computer Science
Cornell University
5149 Upson Hall
Ithaca, NY 14853, USA

Author

Donald E. Knuth
Department of Computer Science, Stanford University
Stanford, CA 94305-2140, USA

CR Subject Classification (1991): I.3.5, F.2.2, G.2.2

1991 Mathematics Subject Classification: 52B55, 52C05, 68Q25, 52B40, 05C20

ISBN 3-540-55611-7 Springer-Verlag Berlin Heidelberg New York
ISBN 0-387-55611-7 Springer-Verlag New York Berlin Heidelberg

Typesetting: Camera ready by author/editor
Printing and binding: Druckhaus Beltz, Hemsbach/Bergstr.
45/3140-543210 - Printed on acid-free paper

Preface

A FEW YEARS AGO some students and I were looking at a map that pinpointed the locations of about 100 cities. We asked ourselves, "Which of these cities are neighbors of each other?" We knew intuitively that some pairs of cities were neighbors and others were not; we wanted to find a formal mathematical characterization that would match our intuition.

Our first solution was rather complicated. We decided to say that points p and q of a given set S are neighbors if the set V_p of all points in the plane that are closer to p than to any other point of S is adjacent to the set V_q of all points that are closest to q. Another way to state this condition is so say that V_p and V_q have a common boundary point t; point t is then equidistant from p and q, and every point between p and t belongs to V_p, while every point between q and t belongs to V_q. After several more minutes we realized that the key fact was the existence of a circle running through points p and q (centered at t), with no points of S inside the circle.

We began to look for an algorithm that would find all neighboring pairs of points $\{p, q\}$, according to our criterion. But time ran out; our meeting had to break up, and we went our separate ways. I wonder what would have happened if we had had more time to explore the problem on our own, before learning that it was a famous problem in computational geometry.

Leo Guibas's office was next to mine, and we soon learned from him that points p and q are neighbors in S by our definition if and only if the line segment pq belongs to the so-called Delaunay triangulation of S. I had never encountered that branch of geometry before, and I hadn't had time to read much of the fast-growing literature of computational geometry. After all, I had never promised to write a book about such things, and the other topics that had kept me going for 30 years were already proving to be more than adequate to occupy an entire lifetime! Furthermore I knew that my geometric intuition was rather poor; algebra and logic have always been much easier for me than visualization. I have absolutely no ability to understand 3-dimensional objects until I have built physical models to represent them.

Leo gave me a reprint of [38], a paper that explains (among other things) how to compute the Delaunay triangulation of n points in $O(n \log n)$ steps, using simple primitive operations and elegant data structures. I knew I didn't have time to read it, but I was immediately fascinated. Here was the clarity I was looking for, a paper that provided a bridge between algebraic and geometric intuition. I still had difficulty, however, understanding some of the proofs, which relied on diagrams that illustrated particular cases. I craved a proof of the algorithm that could be checked by a computer.

So I spent a pleasant afternoon developing a set of five axioms from which algorithms could be devised for a simpler problem—to find the convex hull of points in the plane. These algorithms for convex hull did not depend directly on the coordinates of the points; they relied only on testing whether or not a point p lies to the left or right of a line from q to r, when p, q, and r are given. Moreover, the algorithms could be proved correct purely from the axioms. I didn't need to draw any diagrams, although of course I constructed the proofs with images in mind.

D. E. Knuth

Axioms and Hulls

Springer-Verlag

Berlin Heidelberg New York
London Paris Tokyo
Hong Kong Barcelona
Budapest

A new set of axioms defines a new "universe." We can experience some of the
fun of Star Trek adventurers when we explore the properties of a given list of logical
rules. Of course, many sets of axioms are contradictory or immediately seen to be
trivial; others are too "random" to be interesting. But some sets, like the axioms for
group theory or for projective geometry, lead to immensely fruitful territory. I soon
realized that the five axioms I had stumbled across were leading to a potentially rich
theory, worthy of further study.

I wrote three pages of notes, to show to Leo and to keep on file for future recre-
ation. Then I noticed a few more things, and decided to write a short paper on the
subject. I had learned in the meantime that the mathematical systems defined by my
axioms were equivalent to a special class of oriented matroids, so I began to refresh
my memory of that subject. During a long plane ride to Singapore I played with the
ideas some more, and found that "vortex-free tournaments" have many nice properties intimately related to four axioms of my five. A few days later I proved a theorem
that appears in section 6 of these notes, while sitting next to a lake in Singapore's
exotic Botanical Gardens.

Every time I wrote up one topic, other interesting questions would suggest themselves; one step led naturally to another and then another, often with unexpected ties
to other branches of mathematics and computer science. Soon the manuscript for
my "short paper" had grown to more than 50 pages, but I knew that I was not yet
half done. I had become thoroughly hooked on the subject—constantly aware that
I was supposed to be doing other things, yet unable to resist mathematical beauty.
I was nowhere near a natural boundary at which I could terminate these initial ex-
plorations. After supervising about 30 Ph.D. dissertations, I felt like I was embarking
on a second thesis of my own.

Finally I completed the first 16 sections of the present notes, which are devoted
entirely to axioms for the computation of convex hulls in the plane. I had learned a
lot, but I knew that this was merely a beginning; I had investigated those axioms only
to prepare myself for the real goal, which was to study the more complex problem
of Delaunay triangulation. The latter problem requires axioms for another primitive
operation, namely to test whether a point p lies inside the circle that passes through
three other points $\{q, r, s\}$.

With a mixture of excitement and trepidation, I turned to the Delaunay triangulation problem and found to my great relief that the same five axioms would work for
that problem as well, with only minor extensions. Thus, all of my warmup exercises
had turned out to be immediately applicable to the problem I had hoped to solve.
Hurray! The remaining work went quickly.

I have tried to write these notes in such a way that readers may share the fun I had
during an exhilarating voyage of discovery. Let's face it: Research is thrilling. Instead
of merely presenting the facts, I have tried here to give a reasonably faithful account
of the questions I asked and the answers I liked, following closely the chronological
order in which the work was done. I did not know what would appear in section $n+1$
of these notes when I was exploring the topics discussed in section n. The final section
contains some of the questions that seem to demand answers next. Calls for further
exploration of the terrain can be heard from many directions.

Of course much of this work parallels the research of others; the theory developed here is merely an introduction to the vast and beautiful subject of oriented matroids, which will surely continue to provide inspiration to many more generations of researchers in mathematics and computer science. The most remarkable thing about oriented matroids is perhaps that they are extremely interesting even in the simplest, lowest-dimensional cases.

Section 15 of the present notes can be read separately, because it is largely independent of the rest of the material. It discusses "parsimonious algorithms," a notion that is a natural outgrowth of any axiom-based approach to algorithm design: We say that an algorithm is *parsimonious* if it never makes a test for which the outcome could have been anticipated from the results of previous tests, with respect to a given set of axioms. Algorithms that meet this condition are robust, in instructive ways that may prove to be important in practice, although stronger types of robustness are achievable in the convex hull and Delaunay triangulation problems.

This book may, incidentally, be interesting to typography buffs as well as to computer scientists, because of the rapid turnaround time provided by Springer-Verlag. It is the first publication to use the final revision of the Computer Modern typefaces, released two weeks ago. I made the arrowheads longer and stronger, so that they will not disappear so easily on xerox copies; and I redesigned a few of the letterforms, such as $\mathcal{I}$, $\mathcal{T}$, and δ. There is also an improved method for digitization at low resolution. The new characters do not affect any of the line breaks or page breaks made by TEX, because they fit into exactly the same size boxes as the old ones did. Everyone now using the Computer Modern fonts of 1986 should soon be able to install the 1992 fonts in their place, at little or no cost. I promise not to change them again.

— *Donald E. Knuth*
Stanford, California
April 1992

Summary. A CC system is defined to be a relation on ordered triples of points that satisfy five simple axioms obeyed by the "counterclockwise" relation on points in the plane. A CCC system is a relation on ordered quadruples, satisfying five simple axioms obeyed by the "incircle" relation. In this monograph, the properties of these axioms are developed and related to other abstract notions such as oriented matroids, chirotopes, primitive sorting networks, and arrangements of pseudolines. Decision procedures based on the CC axioms turn out to be NP-complete, although nice characterizations of CC structures are available. Efficient algorithms are presented for finding convex hulls in any CC system and Delaunay triangulations in any CCC system. Practical methods for satisfying the axioms with arbitrarily degenerate data lead to what may well be the best method now known for Delaunay triangulations and Voronoi diagrams in the Euclidean plane. The underlying theme of this work is a philosophy of algorithm design based on simple primitive operations that satisfy clear and concise axioms.

Contents

ONE WAY TO ADVANCE the science of computational geometry is to make a comprehensive study of primitive operations that are used in many different algorithms. This monograph attempts such an investigation in the case of two primitive predicates: The *counterclockwise* relation pqr, which states that the circle through points (p, q, r) is traversed counterclockwise when we encounter the points in cyclic order $p, q, r, p, \ldots$; and the *incircle* relation $pqrs$, which states that s lies inside that circle if pqr is true, or outside that circle if pqr is false.

The counterclockwise and incircle predicates can be applied in many ways. For example, the line segment pq intersects the line segment rs if and only if $pqr \neq pqs$ and $prs \neq qrs$. But the principal applications studied below are the computation of convex hulls (the ordered pairs of points pq such pqr holds for all other points r) and Delaunay triangulations (the ordered triples Δpqr such that $spqr$ holds for all other points s). Delaunay triangulations are of special importance because Voronoi regions are easily calculated once the Delaunay triangulation is known.

The value of an axiomatic approach to geometrical questions has been clear ever since Euclid published his *Elements* about 2300 years ago. Once we know the essential properties of the objects we are dealing with, we can construct rigorous proofs about what is true and what is false. Axioms supplement our geometric intuition, giving us new ways to view a problem via manipulation of logical formulas; we often see patterns in symbols that we cannot see in diagrams, just as we often see patterns in diagrams that we cannot see in symbols. Axioms are especially vital in computational geometry, when an algorithm must make discrete decisions between alternative procedures. Many algorithms that rely on floating-point arithmetic are doomed to failure unless floating-point approximations obey suitable laws. Horror stories abound about methods that "blow up" or produce unacceptable results.

Section 1 introduces five basic axioms for the counterclockwise predicate that will be studied in the remainder of these notes. A set of triples pqr that obeys these axioms will be called a *CC system*. The five axioms are shown to be independent of each other in section 2, which also considers a number of alternative axioms and introduces a graphical technique by which logical manipulations on ternary predicates are easily performed by hand. Section 3 investigates the general systems that arise when only the first four axioms are assumed; then section 4 investigates what happens when we introduce Axiom 5 but omit Axiom 4. The latter systems turn out to be much more interesting, and we call them pre-CC systems. Pre-CC systems are intimately related to a special kind of graph called a *vortex-free tournament*: This is a directed graph in which either $p \to q$ or $q \to p$ holds for all $p \neq q$, with the special property that no four points form a "vortex" (a subtournament consisting of a 3-cycle and a source or sink). Vortex-free tournaments are shown to have a very simple structure, and later sections of the notes demonstrate that this structure is the key property underlying the counterclockwise relation as well as convexity in higher dimensions. Section 5 shows that pre-CC systems contain most of the structure of CC systems. Section 6, on the other hand, establishes a negative result: Although vortex-free tournaments have a very simple structure, they do not have a simple decision procedure; the problem of determining whether a given set of relations is consistent with the existence of a vortex-free tournament turns out to be NP-complete.

Sections 7 and 8 plunge deeper into the internal structure of CC systems, showing that all such systems can be represented conveniently and canonically in terms of *primitive sorting networks*—networks of comparators that had previously been studied in connection with an entirely different problem. This characterization makes it possible to determine the asymptotic number of CC systems, in section 9, where related questions concerning *arrangements of pseudolines* are also considered.

Indeed, CC systems are equivalent to combinatorial structures that have arisen in a variety of different contexts. Section 10 demonstrates a two-to-one correspondence between pre-CC systems and *uniform oriented matroids* of rank 3, which are defined by axioms of a quite different nature.

The study of convex hulls begins in section 11, which presents an efficient method for finding the convex hull of any CC system using only the *pqr* predicate. Another such algorithm, which turns out to have a surprisingly short computer program, is described in section 12. Section 13 compares various implementations of these algorithms by presenting the results of empirical tests on several kinds of data. Section 14 explains how to make the algorithms robust, by avoiding degenerate situations that otherwise occur when points are collinear or coincident.

A weaker form of robustness is the subject of section 15: An algorithm is said to be *parsimonious* if it never evaluates a primitive predicate whose value could have been deduced from previously evaluated predicates. General principles of parsimonious algorithms are discussed in the context of sorting, then a particular algorithm for convex hulls is shown to be parsimonious.

Section 16 wraps up the study of abstract CC systems by noting that CC systems can be composed with each other, using operations analogous to cartesian products.

The incircle predicate is introduced in section 17, where it is shown to obey axioms that are identical to those of CC systems except with another point thrown in. The resulting sets of quadruples *pqrs* are said to form a *CCC system*. The associated theory, which fortunately turns out to be quite elementary, leads to section 18, which might be considered the "grand climax" of this monograph: An efficient algorithm is derived that will find the Delaunay triangulation of any given CCC system. In particular, when the algorithm of section 18 is applied to points in the plane under the usual interpretation of the incircle predicate, it becomes the shortest and fastest method known to the author for computing Delaunay triangulations (hence Voronoi regions). Moreover, the degeneracy-removal techniques discussed in section 19 make this algorithm highly robust. Although the method is quite similar to the randomized incremental procedure of [36], it incorporates several simplifications of practical importance.

Section 20 explains how the two-dimensional concepts of CC and CCC systems fit into more general systems in higher dimensions. Section 21 closes with a historical review of related literature, and section 22 lists several open problems. A number of additional problems are stated throughout the text.

After writing these notes, the author has become ever more convinced of the value of axiomatic methods as a sound basis for computer science in general and for computational geometry in particular. In fact, one of the Delaunay triangulation algorithms he had hoped to prove correct turned out in fact to be invalid, although the method had produced satisfactory results in preliminary tests. Diagrams and special cases are essential for intuition, but they are notoriously poor substitutes for rigorous logic! Once the axioms for CCC systems were understood, it was easy to construct a counterexample to the ill-fated algorithm and (a few hours later) to find a correct procedure. The robust procedure of sections 18 and 19 would probably never have been found without the knowledge gained while preparing sections 1–17.

Disclaimer. This monograph is rather long, because the material seems to be interesting enough to deserve an expository, self-contained treatment. The author has tried to organize things in such a way that the material can be skimmed; for example, cross references between sections are frequently provided. Many of the lemmas and theorems presented below are equivalent to results that are already well known in other formulations. However, references to this other work have often been deferred, in order to avoid interrupting the main flow of ideas. Section 21 attempts to outline the full history of the subject and to give credit where credit is due.

Claimer. On the other hand, the following things appear to be new contributions, as far as the author is aware: the axioms for CC systems in section 1, and in particular the fact that Axiom 5 implies its mirror image; the preliminary investigation of "interior triple systems" of section 3, and in particular the $2^{\Theta(n^3)}$ bound established there; the NP-completeness results of section 6, except for the known material about CSAT; the canonical form of a reflection network in section 8, and the simple proof of (9.5) based on that form; the numerical results in sections 9 and 20, except for the smaller cases; the proof of equivalence between CC systems and oriented matroids in section 10, which seems to be simpler than the treatments of [48] and [51] even when the latter are specialized to uniform matroids of rank 3; the convex hull algorithms of sections 11 and 12; the proof in section 14 that lexicographic order is not sufficient to break ties in degenerate cases; the examples of parsimonious algorithms in section 15; the constructions of section 16; the characterization of incircle in section 17; the general algorithm of section 18, except for the analysis of that algorithm borrowed from [36]; the technique for degeneracy removal in section 19, and the accompanying remarks about calculating signs of determinants via exact computations on floating-point data; the notion of a dual hypertournament (not necessarily geometric) in section 20.

1. Axioms

Let us begin by deriving axioms that hold for the counterclockwise relation between sets of up to five points in the Euclidean plane. If points p, q, r have Cartesian coordinates (x_p, y_p), (x_q, y_q), and (x_r, y_r), it is well known that the counterclockwise predicate corresponds to the sign of a determinant:

$$pqr \iff \det \begin{pmatrix} x_p & y_p & 1 \\ x_q & y_q & 1 \\ x_r & y_r & 1 \end{pmatrix} > 0. \tag{1.1}$$

We shall denote this determinant by $|pqr|$.

The simplest axioms involve only three points:

Axiom 1 (cyclic symmetry). $pqr \implies qrp$.

Axiom 2 (antisymmetry). $pqr \implies \neg prq$.

Axiom 3 (nondegeneracy). $pqr \lor prq$.

In all cases there is an implied quantification "for all distinct points p, q, r in a given set S." Axioms 1 and 2 are simple consequences of the determinant identities $|pqr| = |qrp| = -|prq|$. We assume in Axiom 3 that no three points of S are collinear; collinear points are characterized by the relation $|pqr| = 0$.

The next axiom applies to four distinct points:

Axiom 4 (interiority). $tqr \land ptr \land pqt \implies pqr$.

Intuitively, if t lies to the left of the directed lines qr, rp, and pq, then t must be inside the triangle pqr, which must have counterclockwise orientation. Axiom 4 is a consequence of the determinant identity

$$|pqr| = |tqr| + |ptr| + |pqt|, \tag{1.2}$$

which is a consequence of expanding the left hand side of

$$\det \begin{pmatrix} x_p & y_p & 1 & 1 \\ x_q & y_q & 1 & 1 \\ x_r & y_r & 1 & 1 \\ x_t & y_t & 1 & 1 \end{pmatrix} = 0$$

by cofactors of its last column. We shall write '$t \in \Delta pqr$' as an abbreviation for the three relations '$tqr \land ptr \land pqt$'. These three relations can also be written '$tpq \land tqr \land trp$', by Axiom 1.

Cramer's rule tells us that any four points of a plane satisfy

$$t = \frac{|tqr|}{|pqr|} p + \frac{|ptr|}{|pqr|} q + \frac{|pqt|}{|pqr|} r \tag{1.3}$$

if $|pqr| \neq 0$. Thus, $t \in \Delta pqr$ implies that point t is a convex combination of the points p, q, and r. This property leads to a fifth axiom, which applies to any five distinct points:

Axiom 5 (transitivity). $tsp \land tsq \land tsr \land tpq \land tqr \implies tpr$.

The first three hypotheses state that points p, q, r lie in the halfplane to the left of ts; the last two hypotheses state that q is left of tp and r is left of tq. Hence the conclusion is geometrically obvious that r is left of tp. A formal proof follows from the fact that $\neg tpr$ implies trp, hence $t \in \Delta pqr$, hence t is a convex combination of p, q, r; hence the determinant $|tst|$ is a convex combination of the positive determinants $|tsp|, |tsq|, |tsr|$. But that is impossible because $|tst| = 0$.

The same argument yields a similar axiom:

Axiom 5′ (dual transitivity). $stp \land stq \land str \land tpq \land tqr \implies tpr$.

This statement is not obviously a consequence of Axioms 1–5; but we can in fact deduce it as follows, without even using Axiom 4. Assume that Axiom 5′ fails, so that

$$stp \land stq \land str \land tpq \land tqr \land trp$$

by Axioms 1–3. We can now prove that

$$spq \implies srp,$$

as follows. The implication certainly holds if pqr is true, since Axiom 5 tells us that $pqs \land pqt \land pqr \land pst \land ptr \implies psr$. Suppose $spq \land spr \land prq$; then we must have rqs, otherwise Axiom 5 would say that $rsq \land rsp \land rst \land rqp \land rpt \implies rqt$. But rqs causes another problem, since Axiom 5 also carries the implication $qsr \land qst \land qsp \land qrt \land qtp \implies qrp$.

Thus we have proved $spq \implies srp$ with three applications of Axiom 5. A symmetric argument shows that $srp \implies sqr$, and that $sqr \implies spq$; thus

$$spq \implies srp \implies sqr \implies spq$$

and we have either $spq \land srp \land sqr$ or $sqp \land spr \land srq$. But both of these contradict Axiom 5, since

$$stp \land stq \land str \land spq \land sqr \implies spr\,;$$
$$stq \land stp \land str \land sqp \land spr \implies sqr\,.$$

Our proof that Axioms 1, 2, 3, and 5 imply Axiom 5′ is now complete. A similar proof shows that 1, 2, 3, and 5′ imply 5; we just complement the value of each triple containing s in the argument above.

In later sections of these notes we will see that Axioms 1–5 are strong enough to construct "convex hulls" that have many of the familiar properties of convex hulls in the plane. We will also see that Axioms 1–5 define a theory equivalent to several other theories that have arisen in other contexts. Therefore it is reasonable to give a special name to three-point predicates that satisfy Axioms 1–5; we shall call them *CC systems* (short for "counterclockwise systems").

CC systems do not capture all the properties of counterclockwise relations between points in the plane; determinants satisfy many weird and wonderful identities, such as the syzygy

$$|pqw||rpv||qry||prx||quz| + |rpv||qru||rpz||qpy||qwx| + |qru||pqw||rpz||prx||qvy|$$
$$+ |qru||pqw||rpz||qpy||rvx| + |pqw||rpv||pqx||rqz||ruy| + |rpv||qru||pqx||qpy||rwz|$$
$$+ |rpv||qru||pqx||rqz||pwy| + |qru||pqw||qry||prx||pvz| + |pqw||rpv||qry||rqz||pux|$$
$$= 0\,, \tag{1.4}$$

which is related to the theorem of Pappus. This identity [8] implies that the counter-clockwise triples of any nine points in the plane must satisfy the complicated axiom

$$\neg \big((pqw \oplus rpv \oplus qry \oplus prx \oplus quz) \wedge (rpv \oplus qru \oplus rpz \oplus qpy \oplus qwx)$$

$$\wedge \cdots \wedge (pqw \oplus rpv \oplus qry \oplus rqz \oplus pux) \big), \tag{1.5}$$

where $\oplus$ denotes exclusive or. For if each of the parenthesized clauses in (1.5) were true, each term on the left of (1.4) would be positive, and the sum could not be zero.

Incidentally, identity (1.4) has cyclic symmetry with respect to the transforma-tion $p \to q \to r \to p$, $u \to v \to w \to u$, $x \to y \to z \to x$. It also has a less obvious symmetry in which p, u, x remain fixed and the other elements switch places in their orbits: $q \leftrightarrow r$, $v \leftrightarrow w$, $y \leftrightarrow z$. (In the latter case, the first four determinants in each term change sign.) Thus it has six symmetries altogether.

We will show below that it is possible to construct a CC system such that all of the counterclockwise triples pqw, rpv, qvy, ..., pux occurring in (1.4)and (1.5) are true. Thus CC systems can violate (1.5); they are more general than the systems obtainable from actual points in Euclidean geometry. Indeed, this is not surprising, when we consider that Axioms 1, 2, 3, and 5 were obtained entirely by considering configurations of at most 5 points. We could hardly expect such axioms to be strong enough to deduce the 9-point theorem of Pappus, which states that if eight of the triples of points pux, pvx, pwy, qvy, qwy, quz, rwz, ruz, rvx are collinear, then the ninth triple is also collinear. (A diagram appears in section 7 below.)

If a CC system can arise from actual points in the plane, we will call it *realiz-able*. The fact that unrealizable CC systems exist is no real handicap, because we can construct efficient algorithms that solve geometric problems in any CC system. Such algorithms are more general than algorithms that work only with coordinates of points, and proofs of correctness can be quite simple because CC systems are defined by very simple axioms.

2. Independence

We have seen that Axioms 1, 2, 3, and 5 imply Axiom 5', and it is natural to wonder whether Axioms 1–5 themselves contain some redundancy. It turns out that each of them is independent, in the sense that no four axioms together are strong enough to imply the fifth.

Axiom 2 is clearly independent, since the other axioms hold if pqr is uniformly true for all distinct points p, q, r. Axiom 3 is also independent, since the other axioms hold if pqr is uniformly false.

Axiom 1 is independent because Axioms 2 and 3 are valid on a three-element set $\{a, b, c\}$ for which we have

$$abc, \ \neg acb, \ bac, \ \neg bca, \ cab, \ \neg cba. \tag{2.1}$$

(Axioms 4 and 5 are vacuously satisfied on any three-element set.)

A ternary relation pqr that satisfies Axioms 1–3 is unambiguously specified by exhibiting a triple for each three-element subset. These triples are independent, so there are $2^{\binom{n}{3}}$ ways to satisfy Axioms 1–3 over an n-element set.

A four-element set $\{a, b, c, d\}$ with triples

$$dbc, \ adc, \ abd, \ cba \tag{2.2}$$

and their cyclic shifts satisfies Axiom 1–3 but not Axiom 4; furthermore Axiom 5 holds vacuously. Therefore Axiom 4 is independent.

Finally, we can establish the independence of Axiom 5 by constructing triples on $\{a, b, c, d, e\}$ that satisfy only Axioms 1–4. We will do this by introducing several ideas that will be useful below. Consider the somewhat symmetrical triples

$$abc, \ dab, \ dbc, \ dca, \ eab, \ ebc, \ eca, \ ead, \ ebd, \ ecd. \tag{2.3}$$

For each point p we can form a directed graph (digraph) with arcs $q \to r$ iff pqr holds; the five graphs in this case are

$$
a: \begin{smallmatrix} b \to c \\ \downarrow \!\!\times\!\! \uparrow \\ d \to e \end{smallmatrix}, \quad
b: \begin{smallmatrix} c \to a \\ \downarrow \!\!\times\!\! \uparrow \\ d \to e \end{smallmatrix}, \quad
c: \begin{smallmatrix} a \to b \\ \downarrow \!\!\times\!\! \uparrow \\ d \to e \end{smallmatrix}, \quad
d: \begin{smallmatrix} a \to b \\ \uparrow \!\!\times\!\! \uparrow \\ c \leftarrow e \end{smallmatrix}, \quad
e: \begin{smallmatrix} a \to b \\ \uparrow \!\!\times\!\! \downarrow \\ c \to d \end{smallmatrix}.
$$

In fact, these directed graphs are *tournaments* [56]; that is, either $q \to r$ or $r \to q$ appears, for each pair of distinct vertices q and r. Any system of triples satisfying Axioms 2 and 3 on an n-element set corresponds to a set of n tournaments, and Axiom 1 adds that the arc $q \to r$ appears in the tournament for p if and only if the arcs $r \to p$ and $p \to q$ appear respectively in the tournaments for q and r.

A tournament containing no 3-cycles is called *transitive*. Such tournaments contain no cycles whatever, because any k-cycle $a_1 \to a_2 \to \cdots \to a_k \to a_1$ for $k > 3$ can always be shortened to $a_1 \to a_3 \to \cdots \to a_k \to a_1$ if $a_3 \not\to a_1$. Axiom 4 says that any 3-cycle $p \to q \to r \to p$ in a tournament for t must correspond to a triple pqr, whose effects are recorded in arcs of the tournaments for p, q, r that do not involve vertex t. Axiom 5 says that the tournament for t must not contain four vertices forming a 3-cycle pqr and a source s:

$$
t: \begin{smallmatrix} s \to p \\ \downarrow \!\!\times\!\! \uparrow \\ q \to r \end{smallmatrix}.
$$

Axiom 5′ says, similarly, that no derived tournament should contain a 3-cycle and a sink. (A source dominates all other vertices; a sink is dominated by them.)

The tournaments for a, b, c derived from triples (2.3) are transitive linear orderings: They define the respective linear orderings $b < d < e < c$, $c < d < e < a$, and $a < d < e < b$. But the tournaments for d and e violate Axioms 5 and 5′, respectively. The only 3-cycle present in these tournaments is $a \to b \to c \to a$, and the triple abc does appear in (2.3), so Axiom 4 holds. Thus Axioms 5 and 5′ cannot be derived from Axioms 1–4. (Note that any construction satisfying Axioms 1–3 must satisfy both Axioms 5 and 5′ or neither of them.)

The digraph technique provides a convenient way to work with axioms on small triple systems by hand. Indeed, the author drew literally hundreds of diagrams like the above while writing these notes. Digraphs make it easy to show, for example, that Axioms 1–5 imply the law of "interior transitivity,"

$$t \in \Delta pqr \ \wedge \ s \in \Delta pqt \implies s \in \Delta pqr. \tag{2.4}$$

The hypotheses give us tqr, ptr, pqt, sqt, pst, pqs; and Axiom 4 adds pqr. Five arcs of the tournaments

$$p: \begin{array}{c} q \to r \\ \downarrow \nwarrow \uparrow \\ s \to t \end{array} \qquad q: \begin{array}{c} r \to p \\ \times \uparrow \\ s \leftarrow t \end{array}$$

are therefore known, and Axiom 5 will be violated in p's tournament unless we have $s \to r$; hence psr. Similarly, Axioms 5' in q's tournament implies $r \to s$ and qrs. This completes the proof of (2.4). The status of the remaining triple, either trs or srt, is not implied by the hypotheses.

Notice that the interior transitivity law (2.4) holds vacuously in example (2.3). Therefore it is strictly weaker than the transitive law of Axiom 5; we cannot derive Axiom 5 from Axioms 1–4 and (2.4).

Incidentally, we cannot derive law (2.4) from Axioms 1–4 either. Consider, for example, what happens when we switch three of the triples of (2.3):

$$abc, \ dab, \ dbc, \ dca, \ eab, \ ebc, \ eac, \ eda, \ ebd, \ edc. \tag{2.5}$$

These triples are certainly counter-intuitive if we try to think of them in terms of counterclockwise relations between points; they say that $a \in \Delta edc$, $d \in \Delta abc$, $d \in \Delta ebc$, $e \in \Delta abd$, and that no other interior point relations hold. Therefore they satisfy Axioms 1–4 but not 5, 5', or (2.4).

Rule (2.4) is actually a combination of two implications,

$$t \in \Delta pqr \ \wedge \ s \in \Delta pqt \implies sqr, \tag{2.4a}$$
$$t \in \Delta pqr \ \wedge \ s \in \Delta pqt \implies psr; \tag{2.4b}$$

and we can show that each of them implies the other, in the presence of Axioms 1–4. Suppose, for example, that (2.4a) holds but (2.4b) is false. Then we have tqr, trp, tpq, sqt, pst, pqs, sqr, spr, and Axiom 4 yields rst because $p \in \Delta rst$. But $t \in \Delta rsq$ and $p \in \Delta rst$ and spq contradict (2.4a). Similarly, if (2.4b) holds but (2.4a) is false, we have tqr, trp, tpq, sqt, pst, pqs, qsr, psr, hence $q \in \Delta srt$ and srt; then $t \in \Delta srp$ and $q \in \Delta srt$ and spq contradict (2.4b). On the other hand, (2.4a) and (2.4b) are independent if Axiom 4 is lacking; the triples

$$ebc, aec, abe, dbe, ade, abd, dbc, acd, abd, ced \tag{2.6a}$$

violate (2.4b) when $(p, q, r, s, t) = (a, b, c, d, e)$, but their interior point relations $a \in \Delta cde$, $c \in \Delta aed$, $d \in \Delta abe$, $d \in \Delta ace$, $e \in \Delta abc$, $e \in \Delta cad$ do not lead to any violations of (2.4a). The opposite triples

$$cbe, cea, eba, ebd, eda, dba, cbd, dca, cba, dec \tag{2.6b}$$

satisfy (2.4b) but not (2.4a).

While we're cataloguing consequences of Axioms 1–5, we might as well consider yet another rule. Let $\square pqrs$ stand for the four relations $pqr \wedge qrs \wedge rsp \wedge spq$; thus, $\square pqrs$ means that points (p, q, r, s) define a 4-gon, a convex quadrilateral. Axioms 1, 2, 3, and 5 imply that

$$\square pqrs \ \wedge \ t \in \Delta pqr \ \implies \ \square ptrs. \tag{2.7}$$

For the hypotheses give five arcs of the tournaments

$$
\begin{array}{cc}
q \to r & p \to q \\
p : \ \downarrow\!\!\times\!\!\uparrow \ , & r : \ \uparrow\!\!\times\!\!\uparrow \\
s \quad t & s \quad t
\end{array}
$$

and Axiom 5 will fail in p's tournament unless pts; Axiom 5′ will fail in r's tournament unless rst.

Rule (2.7), like rule (2.4), has two parts

$$\square pqrs \ \wedge \ t \in \Delta pqr \ \implies \ trs; \tag{2.7a}$$
$$\square pqrs \ \wedge \ t \in \Delta pqr \ \implies \ spt. \tag{2.7b}$$

Each of these implies the other. For if, say, (2.7b) holds but not (2.7a), we have pqr, qrs, rsp, spq, tqr, ptr, pqt, tsr, and spt, leaving only the status of qts in doubt. If qts is true we have $\square tspq$ and $r \in \Delta tsp$; but tqr contradicts (2.7b). And if qst we have $\square spqr$ and $t \in \Delta spq$; again, rst contradicts (2.7b). A similar argument, reversing the sense of all triples and renaming variables, derives (2.7b) from (2.7a).

Although our proof of (2.7) was very similar to the proof of (2.4), rule (2.7) is actually stronger than (2.4), yet weaker than Axiom 5. The triples

$$abc, bcd, cda, dab, eab, ebc, eca, eda, edb, edc \tag{2.8}$$

satisfy Axioms 1–4 and (2.4), but (2.7) fails when $(p, q, r, s, t) = (a, b, c, d, e)$. Thus Axioms 1–4 and (2.4) are too weak to imply (2.7). But Axioms 1–4 and (2.7) do imply (2.4); to demonstrate this, we need only verify (2.4a). Suppose $t \in \Delta pqr \wedge s \in \Delta pqt \wedge srq$. Then $q \in \Delta rts$, and Axiom 4 implies rts. We cannot have prs, since $p \in \Delta rst$ would imply rst. Therefore psr; we now have $\square srpq \wedge t \in \Delta srp$. Rule (2.7b) yields qst, a contradiction, completing the proof. Finally, Axioms 1–4 and (2.7) hold in (2.3), so they are not strong enough to imply Axiom 5.

It can be shown that Axioms 1–3 together with (2.4a) and (2.7) imply (2.4b), even when Axiom 4 fails.

Although Axioms 1–5 are independent, they do not give the shortest possible definition of a CC system. It is easy to see that Axioms 1 and 2 can both be deduced from Axiom 3 in conjunction with the rule

$$pqr \ \implies \ \neg qpr. \tag{2.9}$$

3. Interior triple systems

Let's take a minute to study further properties of ternary predicates that satisfy Axioms 1–4 but not necessarily any further laws. We shall call them *interior triple systems*, for want of a better name.

Axiom 4 is rather pleasant because it has many symmetries. Besides an obvious circular symmetry with respect to the cycle (p, q, r), we can recast Axiom 4 in various other ways, such as '$tqr \wedge ptr \wedge \neg pqr \implies \neg pqt$', i.e.,

$$rtq \wedge prq \wedge ptr \implies ptq; \tag{3.1}$$

and this is just like the original statement of Axiom 4, but with r playing the role of t. In fact, the axiom is best written as a clause instead of as an implication:

$$rqt \vee prt \vee qpt \vee pqr. \tag{3.2}$$

This clause form makes it easy to see that the axiom is invariant under all 12 permutations of the alternating group A_4. Therefore if we have any *ordered* set of four points (a, b, c, d), we need not test Axiom 4 for all 4! ways of mapping (p, q, r, t) to (a, b, c, d); we merely need to verify two clauses

$$(abd \vee bcd \vee cad \vee acb) \wedge (bad \vee cbd \vee acd \vee abc). \tag{3.3}$$

A large family of interior triple systems on n points $\{a_1, \ldots, a_n\}$ can be obtained by constructing an arbitrary sequence of permutations $P_3, P_4, \ldots, P_n$, where P_k is a linear ordering of $\{1, 2, \ldots, k-1\}$. Given any such sequence, the set of all triplets $a_i a_j a_k$ where $i < k$, $j < k$, and i precedes j in P_k, defines a system in which Axiom 4 holds. For if $i < j < k < l$, we must have

$$(a_i a_j a_l \vee a_j a_k a_l \vee a_k a_i a_l) \wedge (a_j a_i a_l \vee a_k a_j a_l \vee a_i a_k a_l),$$

and this is stronger than (3.3). The number of such systems is

$$2! \ldots (n-2)! \, (n-1)! = 2^{\Omega(n^2 \log n)}. \tag{3.4}$$

We have seen that Axioms 5 and 5' can be viewed as restrictions on the tournaments defined by the triples containing a given point. Axiom 4 imposes no such restriction. Indeed, an interior triple system can be constructed on $\{a_0, a_1, \ldots, a_n\}$ in which point a_0 is associated with any desired tournament. Given a tournament on $\{a_1, \ldots, a_n\}$ for the triples $a_0 a_i a_j$, suitable triples that do not involve a_0 can be obtained by the construction of the previous paragraph, where we define P_k so that all elements $i < k$ such that $a_k \to a_i$ in the given tournament precede all elements $j < k$ such that $a_j \to a_k$. Then if the given tournament contains a cycle $a_i \to a_j \to a_k \to a_i$, where $k = \max(i, j, k)$, the system will contain the triple $a_i a_j a_k$ as required by Axiom 4.

The total number of interior triple systems is in fact considerably larger than the lower bound in (3.4). If $n = 3m$ we can form $2^{n^3/27}$ such systems on the set $S = \{x_1, \ldots, x_m, y_1, \ldots, y_m, z_1, \ldots, z_m\}$ as follows: Let Q_k be an arbitrary subset of $\{1, \ldots, m\} \times \{1, \ldots, m\}$, for $1 \le k \le m$; there are $(2^{m^2})^m = 2^{n^3/27}$ ways to choose

$Q_1, \ldots, Q_m$, and each of these will lead to an interior triple system. If $a, b, c \in S$ and $a < b < c$, where we assume that

$$x_1 < \cdots < x_m < y_1 < \cdots < y_m < z_1 < \cdots < z_m,$$

then we include the triple abc if and only if $a = x_i$ and $b = y_j$ and $c = z_k$ and $(i, j) \in Q_k$; otherwise we include acb.

Property (3.3) is verified as follows: Let $a < b < c < d$. We cannot have $acd \wedge abc$, because that would imply both $c = y_j$ and $c = z_k$. Hence $cad \vee acb$. Similarly we cannot have $abd \wedge bcd$, which would imply $b = y_j = x_i$. Thus $(cad \vee acb) \wedge (bad \vee cbd)$ must hold, and (3.3) must be true.

The interior triple systems just constructed usually fail to satisfy the interior transitivity axiom (2.4). For example, let $p = z_l$, $q = z_k$, $r = x_h$, $s = x_i$, and $t = y_j$, where $h < i < j < k < l$, $(h, j) \in Q_k$, $(h, j) \notin Q_l$, $(i, j) \notin Q_k$, and $(i, j) \in Q_l$. Then we have $r < s < t < q < p$, and it can be checked that $t \in \Delta pqr$ and $s \in \Delta pqt$; but $s \notin \Delta pqr$. Similarly, the systems based on permutations P_k usually violate (2.4).

These constructions show that interior triple systems are quite plentiful. Such systems need not possess structural properties that make them particularly interesting in studies of geometry. But the special class of *transitive interior triple systems*, which satisfy (2.4) in addition to Axioms 1–4, can perhaps be shown to have a more geometric structure. The interested reader may also wish to determine the asymptotic number of transitive interior systems; is it, for example, $2^{\Omega(n^2 \log n)}$?

Incidentally, it is amusing to try replacing Axiom 4 by the much stronger axiom

$$tpq \wedge tqr \implies pqr. \tag{3.5}$$

A simple induction shows that the only triple systems satisfying Axioms 1–3 and (3.5) are isomorphic to the CC systems obtained from the vertices of n-gons in the plane, defined on points $\{a_1, \ldots, a_n\}$ by

$$a_i a_j a_k \iff (i < j < k) \vee (j < k < i) \vee (k < i < j). \tag{3.6}$$

4.   Vortex-free tournaments

Although sets of triples satisfying Axioms 1–4 do not necessarily have many of the counterclockwise properties of points in the plane, it turns out that Axioms 1, 2, 3, and 5 are enough to guarantee a rich geometrical structure. Let us now take a close look at the significance of Axiom 5 and its dual, Axiom 5′. Ternary relations satisfying Axioms 1, 2, 3, and either 5 or 5′ (therefore both 5 and 5′) will be called *pre-CC systems*.

The tournaments associated with individual points of pre-CC systems are of particular interest because they have a nice characterization. Suppose Axioms 5 and 5′ are expressed in clause form,

$$(tps \vee tqs \vee trs \vee tqp \vee trq \vee tpr)$$
$$\wedge \ (spt \vee sqt \vee srt \vee tqp \vee trq \vee tpr); \tag{4.1}$$

this is the same as

$$\neg (tsp \wedge tsq \wedge tsr \wedge tpq \wedge tqr \wedge trp)$$
$$\wedge \ \neg (stp \wedge stq \wedge str \vee tpq \wedge tqr \wedge trp) . \tag{4.2}$$

Both formulations say that the tournament associated with t is *vortex-free*, i.e., that it contains neither the "in-vortex" nor the "out-vortex,"

$$\text{(4.3)}$$

among its 4-point subtournaments. It follows that if $p_1 \to p_2 \to \cdots \to p_m \to p_1$ is any cycle of the tournament and if q is any other point, then there exist j and k such that $p_j \to q \to p_k$.

The study of vortex-free tournaments is facilitated by the idea of *signed points*, namely the original points $a_1, \ldots, a_n$ and their complements $\bar{a}_1, \ldots, \bar{a}_n$. The original points are said to be positive, and their complements are said to be negative. The operation of changing sign is called *negation*, and we define

$$\bar{\bar{a}} = a \qquad \text{and} \qquad |a| = |\bar{a}| = a . \tag{4.4}$$

The relation $a_i \to a_j$ is now extended to signed points by defining $\bar{a}_j \to a_i$, $a_j \to \bar{a}_i$, and $\bar{a}_i \to \bar{a}_j$ whenever $a_i \to a_j$ holds. Thus, negation of a signed point reverses the directions of all arcs touching it.

Lemma. *A tournament is vortex-free if and only if it can be obtained from a transitive tournament by negating a subset of its points.*

Proof. Negating any point of an in-vortex produces an out-vortex, and conversely. Therefore negation preserves vortex-freeness; any tournament obtained from a transitive tournament by repeated negation must be vortex-free.

Let a be any point of a vortex-free tournament. Negate every point p such that $p \to a$, thereby obtaining a tournament such that $a \to p$ for all $p \neq a$. This new tournament is vortex-free, so it cannot contain any cycles. Therefore it is transitive. □

Corollary. *A tournament on n points is vortex-free if and only if there is a string $\alpha_1 \alpha_2 \ldots \alpha_n$ containing each point or its complement, such that*

$$\alpha_j \to \alpha_k \qquad \text{for} \quad 1 \leq j < k \leq n . \tag{4.5}$$

Moreover, it is possible to construct such a string by examining the direction of only $O(n \log n)$ arcs.

Proof. Condition (4.5) is simply a rephrasing of the lemma, in tems of our notational conventions for signed points.

To construct a suitable string $\alpha_1\alpha_2\ldots\alpha_n$, we may choose α_1 to be any signed point. Then if a partial string $\alpha_1\ldots\alpha_k$ has been constructed representing a vortex-free subtournament on k points for some k with $1 \leq k < n$, let p be any point distinct from $|\alpha_1|,\ldots,|\alpha_k|$, and let $\alpha = p$ or $\bar{p}$ according as $\alpha_1 \to p$ or $p \to \alpha_1$. We know from the lemma that there exists j in the range $1 \leq j \leq k$ such that $\alpha_i \to \alpha$ for $1 \leq i \leq j$ and $\alpha \to \alpha_i$ for $j < i \leq k$; the value of j can be determined by using binary search to examine the direction of at most $\lceil \lg k \rceil$ arcs. This yields a string $\alpha'_1\ldots\alpha'_{k+1} = \alpha_1\ldots\alpha_j\,\alpha\,\alpha_{j+1}\ldots\alpha_k$ that represents a subtournament of $k+1$ points; and the process can therefore continue with k replaced by $k+1$ and $\alpha_1\ldots\alpha_k$ replaced by $\alpha'_1\ldots\alpha'_{k+1}$, until $k = n$. □

The proof of this corollary shows that there are precisely $2n$ strings $\alpha_1\alpha_2\ldots\alpha_n$ that represent a given vortex-free tournament by relation (4.5), since there is one string for each choice of α_1. The rest of the string is then uniquely determined. In fact, the $2n$ possible strings are related to each other in a simple way, because

$$\alpha_1\alpha_2\ldots\alpha_n \quad \text{and} \quad \alpha_2\ldots\alpha_n\,\bar{\alpha}_1$$

define the same vortex-free tournament. The set of all strings representing a given tournament is the set of all n-element substrings of the infinite periodic string

$$\alpha_1\alpha_2\ldots\alpha_n\,\bar{\alpha}_1\bar{\alpha}_2\ldots\bar{\alpha}_n\,\alpha_1\alpha_2\ldots\alpha_n\,\bar{\alpha}_1\,\ldots\,, \tag{4.6}$$

because these are the strings we obtain by repeatedly moving the first element to the end and negating it, and because each of the $2n$ signed points occurs exactly once as the first element of one of these substrings.

This method of representation makes it clear that there are precisely $2^n n!/2n = 2^{n-1}(n-1)!$ ways to define a vortex-free tournament on n labeled points.

We can also count the number of nonisomorphic vortex-free tournaments, because there is one for every equivalence class of boolean strings $\sigma_1\sigma_2\ldots\sigma_n$ under the equivalence relation generated by

$$\sigma_1\sigma_2\ldots\sigma_n \equiv \sigma_2\ldots\sigma_n\bar{\sigma}_1. \tag{4.7}$$

(Here each σ_k is 0 or 1, and $\bar{\sigma} = 1 - \sigma$.) For example, when $n = 5$ the equivalence classes on the 32 boolean strings of length 5 are

$00000 \equiv 00001 \equiv 00011 \equiv 00111 \equiv 01111 \equiv 11111 \equiv 11110 \equiv 11100 \equiv 11000 \equiv 10000$;

$00010 \equiv 00101 \equiv 01011 \equiv 10111 \equiv 01110 \equiv 11101 \equiv 11010 \equiv 10100 \equiv 01000 \equiv 10001$;

$00100 \equiv 01001 \equiv 10011 \equiv 00110 \equiv 01101 \equiv 11011 \equiv 10110 \equiv 01100 \equiv 11001 \equiv 10010$;

$01010 \equiv 10101$.

The 0s correspond to positive variables and the 1s correspond to negative variables in a string $\alpha_1\alpha_2\ldots\alpha_n$ that represents a given tournament.

To see why isomorphism of tournaments corresponds to equivalence of boolean strings, consider for example the vortex-free tournaments on $\{a_1, \ldots, a_5\}$ defined by the strings $a_1 \bar{a}_2 a_3 \bar{a}_4 a_5$ and $\bar{a}_3 a_1 \bar{a}_4 a_2 \bar{a}_5$. The corresponding boolean vectors, 01010 and 10101, are equivalent, so the two tournaments are supposed to be isomorphic. And indeed, the tournament defined by $\bar{a}_3 a_1 \bar{a}_4 a_2 \bar{a}_5$ is also defined by $a_1 \bar{a}_4 a_2 \bar{a}_5 a_3$, so we obtain an isomorphism from the first to the second by mapping $(a_1, a_2, a_3, a_4, a_5)$ to $(a_1, a_4, a_2, a_5, a_3)$.

Conversely, inequivalence of the boolean strings implies nonisomorphism of the tournaments. For example, the tournament defined by $a_1 \bar{a}_2 a_3 \bar{a}_4 a_5$ is not isomorphic to, say, $a_2 \bar{a}_1 \bar{a}_4 a_5 a_3$, whose boolean string is $01100 \not\equiv 01010$. For if the original a_k is mapped to $a_{f(k)}$, we could complement $a_{f(2)}$ and $a_{f(4)}$, getting a transitive tournament in which $a_{f(1)} \to \bar{a}_{f(2)} \to a_{f(3)} \to \bar{a}_{f(4)} \to a_{f(5)}$. The 10 strings $\alpha_1 \ldots \alpha_5$ representing that tournament cannot have the form $a_2 \bar{a}_1 \bar{a}_4 a_5 a_3$, because they correspond only to boolean strings of negation patterns that are equivalent to 01010. This argument has been expressed in terms of a particular example with $n = 5$, but it is perfectly general.

Thus the number $N(n)$ of nonisomorphic vortex-free tournaments on n points can be deduced by counting equivalence classes, and we can solve that problem by slightly extending Macmahon's classic solution to the problem of counting all distinct necklace patterns (see [33, pages 139–141]). We have

$$2nN(n) = \sum_{\sigma_1, \ldots, \sigma_n \in \{0,1\}} \sum_{k=0}^{2n-1} [\sigma_1 \ldots \sigma_n = \sigma_{k+1} \ldots \sigma_{k+n}]$$

$$= \sum_{k=0}^{2n-1} \sum_{\sigma_1, \ldots, \sigma_n \in \{0,1\}} [\sigma_1 \ldots \sigma_n = \sigma_{k+1} \ldots \sigma_{k+n}]$$

when we define $\sigma_{j+n} = \bar{\sigma}_j$ for all $j > 0$; for if we write down $2n$ strings $\alpha'_{k+1} \ldots \alpha'_{k+n}$ for each equivalence class starting with any representative $\sigma'_1 \ldots \sigma'_n$ of that class, $\sigma_1 \ldots \sigma_n$ occurs as often as there are solutions to the string equation $\sigma_1 \ldots \sigma_n = \sigma_{k+1} \ldots \sigma_{k+n}$. (In this formula for $2n N(n)$ we are using Iverson's convention, which evaluates bracketed statements to 0 if they are false, to 1 if they are true; see [46].)

Given any value of k, the inner sum over $\sigma_1, \ldots \sigma_n$ is 0 if $g = \gcd(k, 2n)$ divides n, because the condition $\sigma_j = \sigma_{j+n}$ for all j implies that $\sigma_1 = \sigma_{g+1} = \sigma_{2g+1} = \cdots$, and we know that $\sigma_1 \neq \sigma_{n+1}$. But if g does not divide n, the sum is $2^{g/2}$, because we can vary $\sigma_1, \ldots, \sigma_{g/2}$ independently and set $\sigma_{j+g/2} = \bar{\sigma}_j$ for all j; in this case $g/2$ is an odd divisor of n. Thus if $n = 2^l q$ and q is odd, the nonzero terms occur when $\gcd(k, 2n) = 2n/d$, where d divides q; and in such cases $k = 2nr/d$, where r is relatively prime to d. Hence

$$2nN(n) = \sum_{d \backslash q} 2^{n/d} \sum_{r=0}^{d-1} [\gcd(r, d) = 1]$$

$$= \sum_{\substack{d \backslash n \\ d \text{ odd}}} 2^{n/d} \varphi(d), \tag{4.8}$$

and $N(n)$ is determined. In particular, when n is odd, the number of nonisomorphic vortex-free tournaments on n points is one half the number of distinct necklace patterns of length n that can be formed with two kinds of beads. When n is an odd prime p times a power of 2, the number of nonisomorphic vortex-free tournaments is $(2^n + (p-1)2^{n/p})/2n$.

Notice that the vortex-free tournament defined by $\alpha_1\alpha_2\ldots\alpha_n$ is transitive if and only if the corresponding boolean string is equivalent to $00\ldots0$. This occurs if and only if the boolean string has the form 0^k1^{n-k} or 1^k0^{n-k} for some k, if and only if the string $\alpha_1\alpha_2\ldots\alpha_n\bar\alpha_1$ has exactly one change of sign between adjacent elements.

The theory of tournaments includes numerous results about so-called *score vectors* $(s_1, s_2, \ldots, s_n)$, which are the outdegrees of the points, sorted into nondecreasing order [56]. Vortex-free tournaments on 5 or fewer points are characterized up to isomorphism by their score vectors. But the two tournaments defined by

$$a_1a_2a_3\bar a_4a_5a_6 \quad \text{and} \quad a_6a_5\bar a_4a_3a_2a_1$$

both have the score vector $(1, 2, 2, 3, 3, 4)$; they are anti-isomorphic, but not isomorphic.

J. W. Moon [57] has given another characterization of vortex-free tournaments, which he studied because they are precisely the tournaments whose subtournaments are all either transitive or irreducible. (A tournament is said to be reducible if it has more than one strong component, or equivalently if its vertices can be partitioned into nonempty subsets P and Q with $p \to q$ for all $p \in P$ and $q \in Q$.) There is an integer $m \geq 1$ such that the number of blocks of consecutive elements of the same sign is either $2m$ or $2m-1$ in every string $\alpha_1 \ldots \alpha_n$ that defines a given vortex-free tournament, depending on whether the signs of α_1 and α_n are different or the same. Let us say that a vortex-free tournament belongs to class m if m is that integer; thus, transitive tournaments belong to class 1, and a vortex-free tournament on n vertices may belong to a class whose number is as high as $\lceil n/2 \rceil$. Moon proved (4.8) by showing that the number of nonisomorphic vortex-free tournaments of class m on n vertices is

$$\frac{1}{n} \sum_{\substack{k\backslash n \\ k\backslash(2m-1)}} \varphi(k)\binom{n/k}{(2m-1)/k}, \tag{4.9}$$

then summing on m.

Further characterizations of vortex-freeness were recently discovered by Fred Galvin [23]: A tournament is vortex-free if and only if every subtournament of even order contains an even number of cyclic triples $p \to q \to r \to p$, if and only if every subtournament of order $2m$ has exactly m vertices of score less than m. A tournament contains no out-vortex if and only if every subtournament of order $2m$ has at most m vertices of score less than m, if and only if it can be partitioned into two parts P and Q, where P is vortex-free, Q is transitive, and $p \to q$ for all $p \in P$, $q \in Q$.

We are, however, digressing from our main topic of CC systems and geometry. CC systems, which correspond intuitively to arrangements of points in the plane, satisfy all of the axioms considered above; hence every point p in a CC system on

n points has an associated vortex-free tournament defined by a string $\alpha_1\alpha_2\ldots\alpha_{n-1}$ on the remaining points. This string is, in fact, easy to interpret: It represents the order in which the remaining points are encountered when a straight line through p rotates counterclockwise through $180°$. The positive elements of $\alpha_1\alpha_2\ldots\alpha_{n-1}$ are the points to the left of the initial position of this sweep line; the negative elements are those to the right. If the sweep line is given a suitable direction, the positive elements are all encountered "ahead" of p and the negative ones are all encountered "behind" p. We have the counterclockwise triple $p\alpha_1\alpha_k$ iff α_1 and α_k have the same sign. When the sweep line passes α_1, point $|\alpha_1|$ passes to the other side, and the process continues in the same way on $\alpha_2\ldots\alpha_{n-1}\bar\alpha_1$. The $2(n-1)$ different strings $\alpha_1\alpha_2\ldots\alpha_{n-1}$ that define p's tournament correspond to the different initial positions and orientations of the sweep line.

The lemma and corollary we have proved do not rely on Axiom 1, so vortex-free tournaments characterize all sets of triples that satisfy Axioms 2, 3, 5, and 5'. Let us call these *weak pre-CC systems*. The triples of any weak pre-CC system (hence in particular the triples of any CC system) can be represented efficiently in a computer by an $n \times n$ matrix A such that, for each pair of points $p \neq q$, the value of A_{pq} is the position of q or $\bar q$ in a string $\alpha_{p,1}\ldots\alpha_{p,n-1}$ that defines the tournament associated with p; we also need an $n \times n$ boolean matrix B such that B_{pq} is the sign of q in that string. Then we have

$$
\begin{aligned}
pqr \iff & \left((A_{pq} < A_{pr}) \wedge (B_{pq} = B_{pr})\right) \\
& \vee \left((A_{pq} > A_{pr}) \wedge (B_{pq} \neq B_{pr})\right).
\end{aligned}
\tag{4.10}
$$

In practice, the matrices A and B often make it possible to compute the value of a given relation pqr more quickly than evaluating it directly from its definition, since the definition might require the evaluation of a determinant or the analysis of some other complex criteria. The preprocessing time needed to compute A and B involves only $O(n^2 \log n)$ steps, according to the corollary proved above.

The number of weak pre-CC systems on n labelled points is exactly

$$
\left(2^{n-2}(n-2)!\right)^n = 2^{\Theta(n^2 \log n)},
\tag{4.11}
$$

because this is the number of ways to define n independent vortex-free tournaments on $n-1$ points. This is substantially smaller than the total number $2^{(n-2)(n-1)n/2}$ of triple systems that are required to satisfy only Axioms 2 and 3.

Axiom 1 makes the individual tournaments dependent on each other. If $q \to r$ is present in the tournament associated with p, then $r \to p$ is present in the tournament for q and $p \to q$ is in the tournament for r. Thus, the structure of pre-CC systems is more refined than the structure of systems that are known only to be weakly pre-CC.

5. Pre-CC systems and CC systems

Let us extend the idea of signed points to triples, so that

$$
pqr \iff \neg pq\bar r \iff p\bar q\bar r \iff \neg p\bar qr \iff \bar pqr \iff \neg\bar pq\bar r \iff \bar p\bar qr \iff \neg\bar p\bar qr.
\tag{5.1}
$$

Negating a point in a triple system therefore complements the value of all triples that contain that point.

The following theorem shows that pre-CC systems are not much different from full CC systems; thus, Axiom 5 captures almost all the important properties of Axiom 4:

Theorem. *A set of triples is a pre-CC system if and only if it can be obtained from a CC system by negating a subset of its points.*

Proof. Negating a point preserves Axioms 1, 2, and 3, and it interchanges Axioms 5 and 5′. Therefore negation takes pre-CC systems into pre-CC systems, and any system obtained from a CC system by repeated negation must be pre-CC.

Conversely, let a and b be any points of a pre-CC system. Negate the remaining points p if necessary so that abp holds for all p. We will show that the resulting system is a CC system.

Indeed, Axiom 5 implies that the tournament for a is transitive. It follows that Axiom 4 cannot be violated by four points that include the point a; we cannot have

$$(apq \wedge aqr \wedge arp \wedge rqp) \vee (aqp \wedge arq \wedge apr \wedge pqr)$$

when the tournament for a is transitive. Moreover, any four points $\{p, q, r, t\}$ different from a can be ordered such that $p \to q \to r \to t$ in a's tournament. Suppose the tournament for t is defined by the string $\alpha_1 \ldots \alpha_{n-1}$, where $\alpha_1 = a$. Then p, q, r must occur in this string with a positive sign, because we have tap, taq, and tar. Hence the restriction of the tournament for t to $\{p, q, r\}$ is transitive, and $\{p, q, r, t\}$ cannot violate Axiom 4. □

Our proof of the theorem has, in fact, told us more:

Corollary. *A pre-CC system for which at least one point is associated with a transitive tournament is a CC system.*

We will see in section 11 below that the converse of this corollary is also true: Every CC system with at least three points has at least three points associated with a transitive tournament. Such points may be called *extreme points*, since a point of a realizable CC system is associated with a transitive tournament if and only if it lies on the convex hull.

A *signed bijection* is a one-to-one correspondence from one set of signed points to another that sends $\alpha \mapsto \beta$ if and only if it sends $\bar{\alpha} \mapsto \bar{\beta}$. A *signed permutation* is a signed bijection from a set of signed points to itself. There are $2^n n!$ signed permutations on n elements, because there are $n!$ ways to choose the absolute values of the images and 2^n ways to choose the signs. The group of all signed permutations on n elements is the group of automorphisms of the n-cube, also known as the hyperoctahedral group, sometimes denoted B_n (see, for example, [1]).

Let us say that two pre-CC systems are *preisomorphic* if there is a signed bijection σ that carries one into the other in such a way that pqr holds in the first iff $\sigma(p)\sigma(q)\sigma(r)$ holds in the second. The theorem just proved states that every pre-CC system is preisomorphic to a CC system. Nonisomorphic CC systems can sometimes be preisomorphic; for example, it turns out that there are exactly three isomorphism classes of CC systems on 5 elements, all preisomorphic to each other. It follows that every pre-CC system on 5 elements can be obtained by negation and renaming of points from the CC system that corresponds to the vertices of a pentagon.

It is easy to see which CC systems are preisomorphic to n-gons, as defined in (3.6), because we merely need to determine which points can be negated without violating Axiom 4. Axiom 4 applies to subsets of 4 points, and all 4-element subsystems of an n-gon are equivalent to the vertices of a square. If the four points are (a, b, c, d) in counterclockwise order, the valid triples are abc, bcd, cda, and dab; and it is easy to verify that the 16 possible negations all produce CC systems except when we map (a, b, c, d) into $(\bar{a}, b, \bar{c}, d)$ or $(a, \bar{b}, c, \bar{d})$. Thus we obtain a CC system from an n-gon by negation if and only if the negated vertices are consecutive. The three nonisomorphic systems obtained from a pentagon occur when we negate 0, 1, or 2 consecutive vertices:

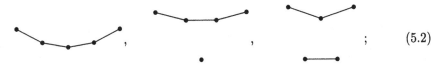

$$; \qquad (5.2)$$

in general when we negate k consecutive vertices of an n-gon, the resulting CC system is equivalent to the sets of points obtained by placing an upward-bending horizontal arrangement of $n - k$ points sufficiently far above a downward-bending arrangement of k points,

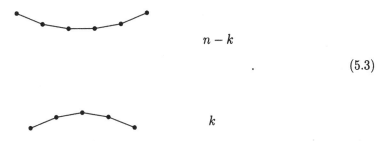

$$. \qquad (5.3)$$

If k and $n - k$ are both greater than 1, the upper and lower points will be concave with respect to each other and the convex hull will be of size 4. Exactly $\lfloor (n + 1)/2 \rfloor$ nonisomorphic CC systems are obtained in this way, because the negation of $n - k$ consecutive points is essentially the same as the negation of k.

An n-gon has exactly $2n$ preautomorphisms (preisomorphisms with itself), generated by the cyclic shift $\sigma = (1, 2, \ldots, n) \mapsto (2, \ldots, n, 1)$ and by the negated reflection $\rho = (1, 2, \ldots, n) \mapsto (\bar{n}, \ldots, \bar{2}, \bar{1})$; the mapping $\sigma \rho \sigma \rho$ is the identity. (Exceptions: When $n = 3$ there are 12 preautomorphisms, generated by σ, ρ, and $(1, 2, 3) \mapsto (1, \bar{2}, \bar{3})$. When $n = 4$, there are 24, generated by σ, ρ, and $(1, 2, 3, 4) \mapsto (1, \bar{3}, \bar{4}, 2)$. A signed permutation that fixes 1 and takes $2 \mapsto k$ or $\bar{k}$ must take $(3, \ldots, n) \mapsto (k + 1, \ldots, n, \bar{2}, \ldots, \overline{k - 1})$ or $(\overline{k + 1}, \ldots, \bar{n}, 2, \ldots, k - 1)$, respectively, in order to preserve the tournament corresponding to 1; this mapping is never a preautomorphism when $n > 4$, so there are no further exceptions.)

If p and q are signed points of a pre-CC system and if p' and q' are signed points of another, there is at most one preisomorphism σ with $\sigma(p) = p'$ and $\sigma(q) = q'$. For if the tournament for p is defined by the string $\alpha_1 \alpha_2 \ldots \alpha_{n-1}$ where $\alpha_1 = q$, and if the tournament for p' is defined by $\alpha_1' \alpha_2' \ldots \alpha_{n-1}'$ where $\alpha_1' = q'$, then we must have $\sigma(\alpha_k) = \alpha_k'$ for all k. Notice that the tournament for p is defined by $\alpha_1 \ldots \alpha_{n-1}$ iff the tournament for $\bar{p}$ is defined by $\alpha_{n-1} \ldots \alpha_1$.

Suppose two CC systems are preisomorphic under the signed bijection σ, and suppose $\sigma(p)$ is positive for all extreme points. In other words, we are assuming that whenever p has a transitive tournament in the first CC system, $\sigma(p)$ is a positive point of the second system. We can prove that $\sigma(p)$ must then be positive for all p. Let $\tau(p) = |\sigma(p)|$ be the ordinary (unsigned) bijection corresponding to σ; if the claim is false, we have $\tau(s) = \overline{\sigma(s)}$ for some s. Let p, q, r be extreme points; then $pqr \iff \tau(p)\tau(q)\tau(r)$, $pqs \iff \neg\tau(p)\tau(q)\tau(s)$, $qrs \iff \neg\tau(q)\tau(r)\tau(s)$, $rps \iff \neg\tau(r)\tau(p)\tau(s)$. Since s is not an extreme point, we can choose p, q, r so that $s \in \Delta pqr$ by letting q and r be the extreme points closest to s in the tournament for p. But then Axiom 4 is violated in the second CC system.

If p is an extreme point of any CC system, we obtain a preisomorphic CC system by negating p (i.e., by mapping $p \mapsto \bar{p}$ and leaving all other points unchanged); this follows from the corollary above, because $\bar{p}$ has a transitive tournament.

Now suppose two CC systems are preisomorphic under σ, and let k be the number of negated points. We call k the *distance* between the two systems under σ. If the original systems are not isomorphic, there must be an extreme point p in the first system for which $\sigma(p)$ is negative. Negating p gives us another CC system whose distance from the second system is only $k - 1$ under σ', where σ' is the mapping $\sigma'(x) = \sigma(x)$ if $|x| \neq p$, $\overline{\sigma(x)}$ if $|x| = p$. Therefore we can go from one CC system to any other preisomorphic CC system by repeatedly negating extreme points.

6. An NP-complete problem

When working with CC systems, we need to deal with partial information—to know when certain sets of triples imply others. Ideally we would like to have an efficient way to solve decision problems involving the vertices of a CC system. But unfortunately it turns out that the axioms, though simple, can lead to situations that are probably very difficult to decide in general. We will prove in this section that it is NP-complete to decide whether specified values of fewer than $\binom{n}{3}$ triples can be completed to a full set of values that satisfies Axioms 1–5. In fact, it turns out to be NP-complete to decide a much simpler problem, which concerns only the $\binom{n-1}{2}$ triples involving a particular point: Given a directed graph, can additional arcs be added to make that graph into a vortex-free tournament?

The analogous question for transitive tournaments is much easier; we can add such arcs if and only if the given digraph has no cycles. But the question of consistency with respect to vortex-freeness is evidently much harder—even though the total number of vortex-free tournaments is only $2^{n-1}(n-1)!$, whose logarithm is asymptotically equal to the log of the number $n!$ of transitive tournaments, and even though problems about cycles are usually "linear" and/or reducible to efficient algorithms based on the theory of matroids.

Before we prove this negative result, a few preparations are necessary. Recall that the satisfiability problem SAT asks if it is possible to find boolean values of variables $(x_1, \ldots, x_n)$ such that every clause in a given set of clauses is true, where each clause has the form $(\sigma_1 \vee \cdots \vee \sigma_k)$, and where each σ_j is either x_i or $\bar{x}_i$, a variable or its complement. The 3SAT problem is the special case where each clause has the form $(\sigma_1 \vee \sigma_2 \vee \sigma_3)$.

We will work with another special case of SAT called CSAT for "complementary satisfiability." In CSAT the clauses come in pairs: Whenever $(\sigma_1 \vee \cdots \vee \sigma_1)$ is a clause, the complementary clause $(\bar{\sigma}_1 \vee \cdots \vee \bar{\sigma}_k)$ is also present. Special cases of CSAT called 3CSAT and 4CSAT involve further restrictions to 3 or 4 literals σ_i per clause.

Lemma. 3SAT *reduces to* 4CSAT.

Proof. Given a set of clauses over n variables $(x_1, \ldots, x_n)$, add a new variable x_0 and construct the new clauses

$$(x_0 \vee \sigma_1 \vee \sigma_2 \vee \sigma_3) \wedge (\bar{x}_0 \vee \bar{\sigma}_1 \vee \bar{\sigma}_2 \vee \bar{\sigma}_3) \tag{6.1}$$

for each clause in the original 3SAT problem. If the original problem is satisfied by the boolean values $x_1 = b_1, \ldots, x_n = b_n$, the new one is satisfied by $x_0 = 0, x_1 = b_1, \ldots, x_n = b_n$. If the new problem is satisfied by $x_0 = b_0, x_1 = b_1, \ldots, x_n = b_n$, the original is satisfied by $x_1 = b_1, \ldots, x_n = b_n$ if $x_0 = 0$, or by $x_1 = \bar{b}_1, \ldots, x_n = \bar{b}_n$ if $x_0 = 1$. ☐

Lemma. 4CSAT *reduces to* 3CSAT.

Proof. Introduce new auxiliary variables a_j for every given pair of clauses $(\sigma_1 \vee \sigma_2 \vee \sigma_3 \vee \sigma_4) \wedge (\bar{\sigma}_1 \vee \bar{\sigma}_2 \vee \bar{\sigma}_3 \vee \bar{\sigma}_4)$, and replace these clauses by

$$(a_j \vee \sigma_1 \vee \sigma_2) \wedge (\bar{a}_j \vee \sigma_3 \vee \sigma_4) \wedge (\bar{a}_j \vee \bar{\sigma}_1 \vee \bar{\sigma}_2) \wedge (a_j \vee \bar{\sigma}_3 \vee \bar{\sigma}_4). \tag{6.2}$$

If the original problem is satisfied by boolean values $(\sigma_1, \sigma_2, \sigma_3, \sigma_4)$, the new clauses are satisfied by taking $a_j = (\bar{\sigma}_1 \wedge \bar{\sigma}_2) \vee (\sigma_3 \wedge \sigma_4)$. Conversely, if the new clauses are satisfied by certain boolean values, we have either $(\sigma_1 \vee \sigma_2) \wedge (\bar{\sigma}_3 \vee \bar{\sigma}_4)$ or $(\bar{\sigma}_1 \vee \bar{\sigma}_2) \wedge (\sigma_3 \vee \sigma_4)$, depending on whether $a_j = 0$ or $a_j = 1$; in both cases the original clauses are satisfied. ☐

Note: The 3CSAT problem is the same as "not-all-equal 3SAT," which is problem LO3 in Garey and Johnson's catalog [24, Appendix 9.1]. This problem was first proved NP-complete by Thomas J. Schaefer [64].

Now we come to the problem of vortex-free completion, VFC, mentioned above: Decide whether a given directed graph is a subgraph of a vortex-free tournament.

Theorem. 3CSAT *reduces to* VFC.

Proof. Given a set of complementary pairs of clauses on $x_1, \ldots, x_n$, we construct a directed graph on the points $p_0, p_1, \ldots, p_n$, plus ten additional points a_j, a'_j, a''_j, b_j,

b'_j, b''_j, cf_j, $'_j$, c''_j, d_j for each clause. If this digraph can be embedded in a vortex-free tournament, the final arc $p_0 \to p_k$ will correspond to the value $x_k = 1$, and $p_k \to p_0$ will correspond to $x_k = 0$. The arcs of the directed graph are obtained by defining 27 arcs corresponding to the jth pair of clauses $(\sigma_1 \vee \sigma_2 \vee \sigma_3) \wedge (\bar{\sigma}_1 \vee \bar{\sigma}_2 \vee \bar{\sigma}_3)$ as follows: If $\sigma_1 = x_k$, include the eight arcs

$$
\begin{array}{ccc}
p_0 & \to a'_j \to & a_j \\
& \times \quad \times & \\
p_k & \to a''_j \to & d_j
\end{array}
\quad ; \tag{6.3}
$$

if $\sigma_1 = \bar{x}_k$, include

$$
\begin{array}{ccc}
p_0 & \to a'_j \to & d_j \\
& \times \quad \times & \\
p_k & \to a''_j \to & a_j
\end{array}
\quad ; \tag{6.4}
$$

Include eight similar arcs for σ_2 using the points b_j, b'_j, b''_j instead of a_j, a'_j, a''_j; and eight more for σ_3, using points c_j, c'_j, c''_j. Also include three additional arcs $a_j \to b_j \to c_j \to a_j$.

Suppose this digraph can be completed to a vortex-free tournament. Then if $p_0 \to p_k$, we must have $a'_j \to a''_j$ and $a_j \to d_j$ if construction (6.3) was used, or $a'_j \to a''_j$ and $d_j \to a_j$ if (6.4) was used. Similarly, the arc $p_k \to p_0$ forces either $d_j \to a_j$ or $a_j \to d_j$, respectively. Thus we have $a_j \to d_j$ if and only if $\sigma_1 = 1$, in our interpretation of the arc direction between p_0 and p_k. The same applies to b_j and c_j. The cycle $a_j \to b_j \to c_j \to a_j$ now means that σ_1, σ_2, and σ_3 cannot be all 0 or all 1. Hence the clauses $(\sigma_1 \vee \sigma_2 \vee \sigma_3) \wedge (\bar{\sigma}_1 \vee \bar{\sigma}_2 \vee \bar{\sigma}_3)$ are satisfied.

Conversely, if all clauses can be satisfied by boolean values $x_1, \ldots, x_n$, we must show that the directed graph can indeed be embedded in a vortex-free tournament. Consistent arcs need to be found between all points, including those between, say, a_j and b''_i for $i \neq j$, without introducing any vortices. A suitable string of signed points to define the desired tournament can be constructed in the form

$$
p_0 \, A_1 \ldots A_m \, \sigma_1 \ldots \sigma_n \, B_1 \ldots B_m \, C_1 \ldots C_m \, d_1 \ldots d_m \tag{6.5}
$$

where $\sigma_k = p_k$ or $\bar{p}_k$ according as $x_k = 1$ or 0, and where (A_j, B_j, C_j) depend on the jth clause pair $(\sigma_1 \vee \sigma_2 \vee \sigma_3) \wedge (\bar{\sigma}_1 \vee \bar{\sigma}_2 \vee \bar{\sigma}_3)$ as follows:

If	then A_j gets	and B_j gets	and C_j gets	
$\sigma_1 = x_k = 1$	a'_j	a_j	a''_j ;	
$\sigma_1 = \bar{x}_k = 1$	a''_j	a_j	a'_j ;	(6.6)
$\sigma_1 = x_k = 0$	a''_j	$\bar{a}_j$	a'_j ;	
$\sigma_1 = \bar{x}_k = 0$	a'_j	$\bar{a}_j$	a''_j .	

A similar construction is applied to σ_2 and σ_3, but with b_j and c_j variables instead of a_j, a'_j, and a''_j. The three variables put into A_j and the three put into C_j can be in any order; but for definiteness we will put a'_j or a''_j first, then b'_j or b''_j, then c'_j or c''_j. The string B_j should be either

$$a_j \bar{c}_j b_j \quad \text{or} \quad b_j \bar{a}_j c_j \quad \text{or} \quad c_j \bar{b}_j a_j$$

$$\text{or} \quad \bar{a}_j c_j \bar{b}_j \quad \text{or} \quad \bar{b}_j a_j \bar{c}_j \quad \text{or} \quad \bar{c}_j b_j \bar{a}_j, \tag{6.7}$$

depending on the variables placed by (6.6) into B_j. Two of a_j, b_j, c_j will have the same sign, and this selects a unique member of (6.7).

For example, if clause-pair j is $(\bar{q} \vee r \vee s) \wedge (q \vee \bar{r} \vee \bar{s})$, and if the clauses are satisfied by $\bar{q} \wedge r \wedge \bar{s}$, then $A_j = a''_j b'_j c''_j$, $B_j = a_j \bar{c}_j b_j$, $C_j = a'_j b''_j c'_j$, and (6.5) will contain

$$p_0 \, a''_j b'_j c''_j \, \bar{q} r s \, a_j \bar{c}_j b_j \, a'_j b''_j c'_j \, d_j$$

when we erase all other points.

It remains to prove that all arcs of the original directed graph are consistent with the arc directions implied by string (6.5). The arcs of (6.3) and (6.4) that touch p_0 and d_j are consistent, because a'_j and a''_j appear between p_0 and d_j in (6.5). The arcs that touch p_k are consistent, because we have either $a'_j a_j a''_j$ or $a''_j \bar{a}_j a'_j$ when $\sigma_1 = x_k$, and we have either $a'_j p_k a''_j$ or $a''_j \bar{p}_k a'_j$ depending on whether $x_k = 1$ or 0. The arcs that touch a_j are consistent, because we have either $a''_j a_j a'_j$ or $a'_j \bar{a}_j a''_j$ when $\sigma_1 = \bar{x}_k$. The same observations apply to b_j and c_j variables. Finally, the arcs $a_j \rightarrow b_j \rightarrow c_j \rightarrow a_j$ are consistent with all of the possibilities in (6.7). $\quad\square$

Corollary. *The problem of deciding whether the values of a given set of triples are consistent with Axioms 1–5 is NP-complete.*

Proof. This problem is clearly in NP. The theorem shows that it is NP-hard just to decide whether triples all involving a single point t will satisfy Axioms 5 and 5'. To complete the proof, we need to show that any vortex-free tournament is the tournament associated with a point of some CC system.

Given any vortex-free tournament on $\{a_1, \ldots, a_n\}$, let a_0 be another point, and define $a_0 \rightarrow a_k$ for all k. Then let $a_i a_j a_k$ be true if and only if at least two of the relations $a_i \rightarrow a_j$, $a_j \rightarrow a_k$, $a_k \rightarrow a_i$ are true. This rule defines a system of triples in which the original tournament is the tournament associated with a_0. We claim that it is, in fact, a CC system. Axioms 1–3 certainly hold.

Suppose $t \in \Delta pqr$; that is, suppose we have tpq, tqr, and trp. Then the inequalities

$$[t \rightarrow p] + [p \rightarrow q] + [q \rightarrow t] \geq 2$$
$$[t \rightarrow q] + [q \rightarrow r] + [r \rightarrow t] \geq 2$$
$$[t \rightarrow r] + [r \rightarrow p] + [p \rightarrow t] \geq 2$$

can be added to give

$$[p \rightarrow q] + [q \rightarrow r] + [r \rightarrow p] + 3 \geq 6.$$

Hence $p \rightarrow q \rightarrow r \rightarrow p$, and Axiom 4 has been verified.

Moreover, the given system has the property that when tpq and $p \to q$ then $t \to p \iff t \to q$. Therefore if $t \in \Delta pqr$ we have either

$$t \to p \wedge t \to q \wedge t \to r$$

or

$$p \to t \wedge q \to t \wedge r \to t.$$

But that is impossible in a vortex-free tournament; so $t \in \Delta pqr$ can occur only when t is the special point a_0. Axiom 5 now follows immediately.

Alternatively, we can use a geometric argument to show that the stated triples form a CC system. Let the given vortex-free tournament be defined by the string of signed points $\alpha_1 \ldots \alpha_n$ and consider arbitrary angles

$$0 < \theta_1 < \cdots < \theta_n < \pi.$$

If $\alpha_j = a_k$ let a_k be the complex number $e^{i\theta j}$; if $\alpha_j = \bar{a}_k$ let $a_k = -e^{i\theta j}$. Let $a_0 = 0$. Then $a_i a_j a_k$ is true as defined above if and only if the points (a_i, a_j, a_k) form a counterclockwise triple in the complex plane. This argument shows that it is NP-hard to determine whether or not a given set of triples is part of a *realizable* CC system. (Indeed, the latter problem may not even be in NP, although Tarski's decision procedure [61] shows that realizability can be tested in finite time.) ☐

7. Fitting tournaments together

Suppose we want to generate, or to imagine that we could generate, all pre-CC systems on seven points $\{0, 1, 2, 3, 4, 5, 6\}$. (We might as well name the points by using the digits themselves, instead of wasting time writing $\{a_0, a_1, \ldots, a_6\}$.) We can assume that the vortex-free tournament associated with 0 is defined by the string 123456, because a signed permutation on $\{2, 3, 4, 5, 6\}$ will produce all other cases. The corollary in section 5 now tells us that our pre-CC system will in fact be a bona fide CC system: Axiom 4 will automatically be satisfied, since the tournament for 0 is transitive.

Let us now proceed to consider all possible tournaments associated with 1. If 1's tournament is defined by a string $\alpha_1 \alpha_2 \alpha_3 \alpha_4 \alpha_5 \alpha_6$ ending with $\alpha_6 = 0$, we see that α_1, α_2, α_3, α_4, and α_5 must be positive; this follows because 012, 013, 014, 015, and 016 are all true. Thus $\alpha_1 \alpha_2 \alpha_3 \alpha_4 \alpha_5$ is a permutation of $\{2, 3, 4, 5, 6\}$. A moment's thought shows that all 5! permutations are possible; we can construct examples in the plane where the points 2, 3, 4, 5, 6 are "seen" from 1 in any desired order, given their counterclockwise order as seen from 0. Suppose, then, that we say the tournament associated with 1 is defined by some string such as 436250.

The tournament associated with 6 can similarly be defined by a string

$$\beta_1 \beta_2 \beta_3 \beta_4 \beta_5 \beta_6$$

that starts with $\beta_1 = 0$, and $\beta_2 \ldots \beta_6$ must be a permutation of the positive points $\{1, 2, 3, 4, 5\}$. Now, however, the permutation is no longer arbitrary. For example, everything preceding 6 in $\alpha_1 \ldots \alpha_5$ must follow 1 in $\beta_2 \ldots \beta_6$, because $1x6$ holds if $61x$. Everything following 6 in $\alpha_1 \ldots \alpha_5$ must also precede 1 in $\beta_2 \ldots \beta_6$.

Further restrictions are present as well. For example, we will not be able to complete the construction if $\beta_2 \ldots \beta_6 = 25143$. The tournament associated with 3 would then include

$$0 \to 1$$
$$\uparrow \;\nwarrow\!\!\!\!\!\diagup\; \uparrow \;,$$
$$4 \to 6$$

violating Axiom 5. Also the tournament associated with 2 would include

$$0 \to 1$$
$$\uparrow \;\nwarrow\!\!\!\!\!\diagup\; \downarrow \;,$$
$$5 \to 6$$

another violation. If $\alpha_1 \ldots \alpha_5 = 43625$, it turns out that the only possibility for $\beta_2 \ldots \beta_6$ is 52134.

Pursuing this line of reasoning, we will discover that the strings defining tournaments for 0, 1, and n in a pre-CC system can be respectively

$$1\,2 \ldots n, \quad \alpha\,n\,\beta^{\mathrm{R}}\,0, \text{ and } 0\,\beta'^{\mathrm{R}}\,1\,\alpha'$$

(where β^{R} is the reverse of string β) if and only if α and α' are permutations on a subset of $\{2, \ldots, n-1\}$ having no inversions in common, and β and β' are permutations on the complementary subset having no inversions in common. (An inversion is a pair of numbers $j < k$ that appears with k to the left of j. In our example above, α and α' both contained the inversion $4\,3$; β and β' both contained $5\,2$.)

This is a rather strong condition, because we can prove without difficulty that the probability for two random permutations to contain no common inversions is at most $(n+1)/2^n$. (This is the probability that each element has at most as many inversions in one permutation as in the reverse of the other.)

The conditions for four or more strings defining compatible tournaments are increasingly complex and restrictive. It appears that we cannot construct nearly as many pre-CC systems as we might have expected, given our study of weak pre-CC systems.

Although it is difficult to piece tournaments together one by one in this manner, there is a fairly simple way to avoid such complications if we try to construct the tournaments in parallel. Instead of thinking of a single directed line that sweeps around one vertex at a time, let us imagine a family of parallel lines, one passing through each point, each directed consistently. If these lines revolve at the same rate, the moment when point p enters into the tournament for q will be the same as the moment when $\bar{q}$ enters the tournament for p; this occurs when the lines through p and q cross, with p visible in the positive direction from q and q visible in the negative direction from p. We can represent this situation by writing $\frac{p}{q}$.

As the parallel lines sweep through $180°$, each pair of points $\{p, q\}$ will be encountered exactly once, either in the form $\frac{p}{q}$ or $\frac{q}{p}$. From these $\binom{n}{2}$ ordered pairs, we can write down strings defining the vortex-free tournaments associated with each point as before, appending p to string q and $\bar{q}$ to string p when the pair $\frac{p}{q}$ appears.

Of course, not every arrangement of ordered pairs will work; we want to define a pre-CC system, not just a weak pre-CC system. Thus the tournament for p must contain the arc $q \to r$ iff the tournament for q contains $p \to q$. The three ordered pairs involving $\{p, q, r\}$ will have at least one variable (say p) occurring both on top and on the bottom, say as $\begin{smallmatrix}p\\r\end{smallmatrix} \begin{smallmatrix}q\\p\end{smallmatrix}$. Then p's tournament will contain $q \to r$; so the tournament for q will be consistent only if $\begin{smallmatrix}q\\r\end{smallmatrix}$ precedes $\begin{smallmatrix}q\\p\end{smallmatrix}$ or $\begin{smallmatrix}r\\q\end{smallmatrix}$ follows $\begin{smallmatrix}p\\q\end{smallmatrix}$, and the tournament for r will be consistent only if $\begin{smallmatrix}r\\q\end{smallmatrix}$ precedes $\begin{smallmatrix}r\\p\end{smallmatrix}$ or $\begin{smallmatrix}q\\r\end{smallmatrix}$ follows $\begin{smallmatrix}p\\r\end{smallmatrix}$. The only way to make both of them consistent is to have $\begin{smallmatrix}q\\r\end{smallmatrix}$ between $\begin{smallmatrix}p\\r\end{smallmatrix}$ and $\begin{smallmatrix}q\\p\end{smallmatrix}$. Similarly, if the pairs involving p are $\begin{smallmatrix}r\\p\end{smallmatrix} \begin{smallmatrix}p\\q\end{smallmatrix}$, we must have $\begin{smallmatrix}r\\q\end{smallmatrix}$ between them.

We have therefore demonstrated the necessity of the following *betweenness rule*, if $\binom{n}{2}$ ordered pairs are supposed to define a pre-CC system:

$$\text{if } \begin{smallmatrix}p\\q\end{smallmatrix} \text{ and } \begin{smallmatrix}r\\p\end{smallmatrix} \text{ occur (in either order),}$$

$$\text{then } \begin{smallmatrix}r\\q\end{smallmatrix} \text{ occurs between them.} \qquad (7.1)$$

Conversely, if an arrangement of $\binom{n}{2}$ ordered pairs obeys the betweenness rule, they define n strings for vortex-free tournaments in which all triples pqr, qrp, rpq have the same value. Therefore they define a pre-CC system.

In fact, they define a CC system. Given an arrangement of $\binom{n}{2}$ ordered pairs satisfying (7.1), let's say that $p \succ q$ if $\begin{smallmatrix}p\\q\end{smallmatrix}$ appears. Then $r \succ p$ and $p \succ q$ implies $r \succ q$, so the relation is transitive. The points can therefore be listed in order $(p_1, p_2, \ldots, p_n)$ so that

$$p_j \succ p_k \quad \Longleftrightarrow \quad j > k. \qquad (7.2)$$

Point p_1 occurs only in the lower row, so its tournament is defined by a string with no negated entries. Thus p_1 has a transitive tournament, and we know from the corollary in section 5 that this guarantees a CC system.

We are now ready to complete an investigation we began in section 1 above: We wish to construct a CC system on nine points that is unrealizable in the plane, by constructing a CC system in which each of the triples occurring in the determinants of identity (1.4) is a counterclockwise triple of points.

Figure 1 on the next page shows a symmetrical configuration of nine points that would correspond to the theorem of Pappus if the lines xv, yw, and zu were straightened so that the triangles now containing p, q, and r shrink to points, and if p, q, r move to those triple-intersection points. We have perturbed p, q, r slightly, and bent three of the lines, so that the triples pux, ruy, qvy, and six others obtained by cyclic rotation

$$p \to q \to r \to p, \qquad u \to v \to w \to u, \qquad x \to y \to z \to x \qquad (7.3)$$

will all have counterclockwise orientation. (Some sort of perturbation and line-bending is obviously necessary if we are to have a diagram, because we know that no CC system containing the triples of (1.4) can be realizable.) The other counter-clockwise triples needed, namely pqw, pqx, qpy, and their cyclic counterparts under (7.3), are clearly present in Figure 1.

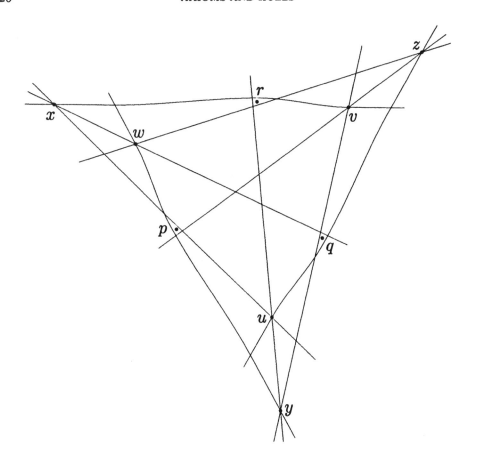

Figure 1. A non-Euclidean CC system.

We can write down $\binom{9}{2} = 36$ ordered pairs by looking at the diagram and imagining parallel sweep lines that rotate counterclockwise after initially pointing directly upward between yu and yv:

$$\begin{smallmatrix} r & u & p & w & w & r & w & x & x & x & p & w & x & x & x & x & r & p & w & x & r & w & w & p & p & v & q & u & u & p & u & y & y & y & q & u \\ y & y & y & y & p & q & u & y & p & u & u & q & q & w & r & v & v & q & v & z & z & z & r & v & z & z & z & z & q & r & v & z & q & v & v & r \end{smallmatrix}. \quad (7.4)$$

It is easy to check that the betweenness condition (7.1) is satisfied; therefore sequence (7.4) defines a CC system. The induced linear ordering (7.2) is

$$z\, v\, q\, y\, r\, u\, p\, w\, x. \qquad (7.5)$$

The tournaments for p, u, and x are

$$p:\ \bar{y}\, w\, x\, \bar{u}\, \bar{q}\, \bar{v}\, \bar{z}\, \bar{r}$$

$$u:\ \bar{y}\, w\, x\, p\, \bar{z}\, \bar{q}\, \bar{v}\, \bar{r}$$

$$x:\ \bar{y}\, \bar{p}\, \bar{u}\, \bar{q}\, \bar{w}\, \bar{r}\, \bar{v}\, \bar{z}, \qquad (7.6)$$

and those for q, r, v, w, y, z are obtained by applying the symmetry relation (7.3). (Actually we get

$$q: \ r\,w\,x\,p\,\bar{z}\,u\,y\,\bar{v}\,;$$

but $r\,w\,x\,p\,\bar{z}\,u\,y\,\bar{v}$ is equivalent to the string $\bar{z}\,u\,y\,\bar{v}\,\bar{r}\,\bar{w}\,\bar{x}\,\bar{p}$ that we get by applying (7.3) to p's string $\bar{y}\,w\,x\,\bar{u}\,\bar{q}\,\bar{v}\,\bar{z}\,\bar{r}$.)

Notice that the first 12 pairs of (7.4), which cover $60°$ of the sweeping process in Figure 1, are transformed into the next 12 pairs by applying (7.3) and flipping. If we keep on going after $180°$, the next pairs will be $\frac{y}{r}\,\frac{y}{u}\,\frac{y}{p}\,\frac{y}{w}\,\frac{p}{w}\ \cdots$, representing (7.4) but flipped. This, in fact, is a general principle that applies to every consistent arrangement of $\binom{n}{2}$ pairs, symmetrical or not: We can always continue to append more pairs by repeating the original $\binom{n}{2}$ pairs upside down, and then by starting the whole cycle again. The consistency condition (7.1) will be satisfied throughout the entire infinite sequence of pairs obtained in this way; for if say the pattern in the first half cycle is $\frac{p}{q}\cdots\frac{r}{q}\cdots\frac{r}{p}$, the infinite sequence includes the pairs

$$\frac{p}{q}\cdots\frac{r}{q}\cdots\frac{r}{p}\cdots\frac{q}{p}\cdots\frac{q}{r}\cdots\frac{p}{r}\cdots\frac{p}{q}\cdots\frac{r}{q}\cdots\frac{r}{p}\cdots\frac{q}{p}\cdots\frac{q}{r}\cdots\frac{p}{r}\cdots\frac{p}{q}\cdots\ . \tag{7.7}$$

If we take any $\binom{n}{2}$ consecutive pairs of the infinite sequence, we get a consistent arrangement that defines an identical sequence of tournaments, because the strings for each tournament are each being shifted and complemented.

We have seen on intuitive grounds (via parallel sweep lines) that the counter-clockwise triples of every realizable CC system can be defined by an arrangement of $\binom{n}{2}$ ordered pairs satisfying the betweenness condition (7.1); we have also proved that every such arrangement defines a CC system. To complete the chain of reasoning, we now want to show that every CC system, realizable or not, is defined by such an arrangement.

Suppose we are given n vortex-free tournaments associated with points labeled $\{1,\ldots,n\}$, where the string α_p defining the tournament for p contains the points $\{1,\ldots,p-1,p+1,\ldots,n\}$. The tournaments are assumed to be consistent; i.e., if $q \to r$ in α_p, then $r \to p$ in α_q and $p \to q$ in α_r. Each string α_p can be represented as a sequence of ordered pairs, using $\frac{p}{q}$ for $\bar{q}$ and $\frac{q}{p}$ for q. We will show how to construct an arrangement of all $\binom{n}{2}$ pairs, containing $\alpha_1, \alpha_2, \ldots, \alpha_n$ as subarrangements. The construction proceeds by induction on n: First we delete 'n' from $\alpha_1, \ldots, \alpha_{n-1}$ and arrange the $\binom{n-1}{2}$ pairs $\frac{p}{q}$ for $1 \le q < p < n$ in some manner consistent with the remainder of $\alpha_1, \ldots, \alpha_{n-1}$. Then we divide the pairs into two classes, assigning $\frac{p}{q}$ to class L if $\bar{p}$ follows $\bar{q}$ in α_n and to class R if $\bar{p}$ precedes $\bar{q}$ in α_n. Notice that $\frac{p}{q}$ is in R iff npq is true iff p follows n in α_q iff $\bar{q}$ follows n in α_p. Therefore we will be done if all pairs of L precede all pairs of R; the pairs of α_n can then all be inserted between L and R.

If the construction runs into trouble, there must be a pair $\frac{p}{q}$ of R immediately followed by a pair $\frac{r}{s}$ of L. We can interchange those pairs if p, q, r, s are distinct, obtaining an arrangement with one less problematic R before L, because the new arrangement will still be consistent with $\alpha_1, \ldots, \alpha_{n-1}$. If p, q, r, s are not distinct, suppose $p = r$. Then npq is true and nps is false, so $\bar{q}$ follows n and n follows $\bar{s}$ in α_p.

Therefore $\bar{q}$ follows $\bar{s}$ in α_p, contradicting the fact that $\frac{p}{q}$ precedes $\frac{p}{s}$. Similarly, if $q = s$ we reach a contradiction after noting that p would have to follow n and n would have to follow r in α_q. The only other possibilities are $p = s$ or $q = r$; but these violate the betweenness condition (7.1), so such an arrangement cannot be consistent with $\alpha_1, \ldots, \alpha_{n-1}$. We have proved that the construction will eventually succeed, after possibly interchanging pairs that don't overlap.

A small example should help make this construction clear. Suppose

$$\alpha_1 = 3\,4\,2\,5, \quad \alpha_2 = 3\,4\,\bar{1}\,5, \quad \alpha_3 = \bar{2}\,\bar{1}\,5\,4, \quad \alpha_4 = \bar{2}\,\bar{1}\,5\,\bar{3}, \quad \alpha_5 = \bar{1}\,\bar{2}\,\bar{4}\,\bar{3}.$$

We begin by setting $n = 2$ and suppressing all entries that are at most 2, trivially obtaining $\frac{2}{1}$. Then n advances to 3, and we obtain $\frac{3}{2}\,\frac{3}{1}\,\frac{2}{1}$ since $\frac{2}{1} \in R$. Then n advances to 4; now we have $\frac{3}{2}$ and $\frac{3}{1} \in L$ and $\frac{2}{1} \in R$, so we obtain the sequence

$$\frac{3}{2}\,\frac{3}{1}\,\frac{4}{2}\,\frac{4}{1}\,\frac{4}{3}\,\frac{2}{1}\,.$$

Finally, when $n = 5$, all pairs except $\frac{4}{3}$ are in L, so we interchange $\frac{4}{3}$ with $\frac{2}{1}$ and obtain

$$\frac{3}{2}\,\frac{3}{1}\,\frac{4}{2}\,\frac{4}{1}\,\frac{2}{1}\,\frac{5}{1}\,\frac{5}{2}\,\frac{5}{4}\,\frac{5}{3}\,\frac{4}{3}\,.$$

Theorem. *Every arrangement of the $\binom{n}{2}$ distinct ordered pairs $\frac{p}{q}$ of n given points as an ordered list satisfying the betweenness condition (7.1) defines a CC system. Conversely, every CC system can be defined by such an arrangement.*

Proof. We have already proved the first part. To prove the converse, we use the fact that any CC system has a point whose tournament is transitive; this will be proved in section 11 below. Thus we can number the points $\{1, \ldots, n\}$ so that the tournament associated with 1 is defined by the string $\alpha_1 = 2\,3 \ldots n$, and the tournaments associated with $2, \ldots, n$ are defined by strings $\alpha_2, \ldots, \alpha_n$ beginning with $\bar{1}$. It follows that the elements of α_p are $\{\bar{1}, \ldots, \overline{p-1}, p+1, \ldots, n\}$, and the construction above can be used to produce the desired arrangement. □

Three distinct points $\{p, q, r\}$ of a CC system are said to allow *mutation* if we cannot deduce the value of pqr from the values of all other triples; they allow *premutation* if complementing pqr yields a pre-CC system. The latter condition is easily recognizable if we look at the associated tournaments: Let us say that q is *adjacent* to r in the tournament for p if q or $\bar{q}$ appears next to r in the infinite string (4.6) representing that tournament; then points p, q, r allow premutation if and only if each is adjacent to the other in the tournament for the third. For example, if q is not adjacent to r in the tournament for p, there are signed points s and t such that p's tournament restricted to $\{q, r, s, t\}$ has the form

$$p: \; q\,s\,r\,t \quad \text{or} \quad p: \; q\,s\,\bar{r}\,t;$$

but then the value of pqr is deducible from the values of pqs, pqt, psr, pst, and prt, by Axiom 5. Conversely, if the tournament for p has the form $p: \; q\,r\,\alpha$ or $p: \; q\,\bar{r}\,\alpha$, for some string α, we can complement pqr without altering any other triples involving p by changing that tournament to $p : r\,q\,\alpha$ or $p : \bar{r}\,q\,\alpha$, respectively. If all three adjaciencies hold, we obtain compatible tournaments when pqr is complemented, so $\{p, q, r\}$ must allow premutation.

Several mutations are easy to spot in (7.4); indeed, it's not difficult to prove that a subsequence like $\begin{smallmatrix} x & x & p \\ p & u & u \end{smallmatrix}$ guarantees mutability. Nine such mutations are possible, one for each of the nine lines in Figure 1.

The triple $\{x, y, z\}$ allows premutation but not mutation, because xyz can be complemented without violating Axiom 5 but not without violating Axiom 4. It is interesting to note that all ten premutations allowed by Figure 1 are disjoint, i.e., independent of each other; we obtain 2^{10} pre-CC systems by assigning arbitrary values to the triples pux, pvz, pwy, qvy, qwx, quz, rwz, ruy, rvx, and xyz. And we obtain 2^9 CC systems by keeping xyz true while assigning arbitrary values to the other nine. Of course many of these systems will be isomorphic. All of the resulting CC systems turn out to be realizable, except the one we began with.

8. Reflection networks

The arrangements in the preceding theorem are strongly related to configurations called "primitive sorting networks" that have arisen in a completely different context (see [45, exercise 5.3.4–36]). A *comparator* $[i : j]$ operates on a sequence of numbers $(x_1, \ldots, x_n)$ by replacing x_i and x_j respectively by $\min(x_i, x_j)$ and $\max(x_i, x_j)$. A *sorting network* is a sequence of comparators that will sort any given sequence $(x_1, \ldots, x_n)$; that is, the successive comparators will produce an output sequence that always satisfies $x_1 \leq \cdots \leq x_n$. A sorting network is called *primitive* if its comparators all have the form $[i : i + 1]$, thus operating on adjacent elements. Floyd proved [20] that a sequence of such comparators is a sorting network if and only if it sorts the single permutation $(n, \ldots, 2, 1)$. We may assume that a sorting network contains no redundant comparators, i.e., no comparators that can be removed because previous comparators ensure that $x_i \leq x_j$. Floyd's theorem implies that an irredundant primitive sorting network is equivalent to a sequence of adjacent *transpositions* $[i, i + 1]$, which changes an array $(x_1, x_2, \ldots, x_n)$ into its reflection $(x_n, \ldots, x_2, x_1)$. We shall call such a sequence a *reflection network* for convenience.

For example, here are the reflection networks that arise for $n = 5$ as a consequence of classical sorting methods called "bubblesort," "cocktail-shaker sort," and "odd-even transposition sort" [45]:

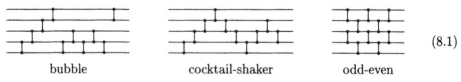

$$\tag{8.1}$$

| bubble | cocktail-shaker | odd-even |

(The reflection properties of the odd-even transposition method have been well known for almost 300 years in the English art of change-ringing; they are the first changes of "Plain Bob" [58, pages 346 and 379].)

A reflection network for n elements consists of exactly $\binom{n}{2}$ transpositions, because every adjacent transposition decreases the number of inversions by 1 if we begin with the array $(n, \ldots, 2, 1)$; this array has $\binom{n}{2}$ inversions, and the final array $(1, 2, \ldots, n)$ has none. We can construct reflection networks easily by starting with $(n, \ldots, 2, 1)$ and repeatedly exchanging any two adjacent elements that happen to be out of order; after $\binom{n}{2}$ steps we will surely arrive at $(1, 2, \ldots, n)$, and the sequence of operations performed will be a reflection network.

We have observed that every arrangement of $\binom{n}{2}$ pairs that satisfies the betweenness rule (7.1) defines a linear order (7.2). In fact, reflection networks are in one-to-one correspondence with betweenness arrangements having a given linear order. If we number the points 1 to n according to the linear order, then the arrangement specifies a sequence of adjacent interchanges that converts $(n, \ldots, 2, 1)$ into $(1, 2, \ldots, n)$. For if the first pair is $\frac{p}{q}$, we must have $p = q + 1$; otherwise there would be an r such that $\frac{p}{r}$ and $\frac{r}{q}$ both appear, without $\frac{p}{q}$ between them. Therefore interchanging p with q is an adjacent interchange in $(n, \ldots, 2, 1)$. Removing $\frac{p}{q}$ from the left, placing $\frac{q}{p}$ at the right, and interchanging the labels of p and q allows us to repeat this argument $\binom{n}{2}$ times. For example, the reflection network corresponding to the 36 pairs of the near-Pappus unrealizable CC system of (7.4) is

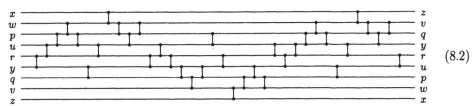

$$(8.2)$$

Conversely, a reflection network defines betweenness arrangements. For if $r > p > q$, the inversion rq must be removed between the times when the inversions rp and pq are removed. For example, the bubblesort network of (8.1) corresponds to the arrangement

$$\begin{smallmatrix} 2 & 3 & 4 & 5 & 3 & 4 & 5 & 4 & 5 & 5 \\ 1 & 1 & 1 & 1 & 2 & 2 & 2 & 3 & 3 & 4 \end{smallmatrix};$$

this, in turn, defines the vortex-free tournaments

$$1:2345, \quad 2:\bar{1}345, \quad 3:\bar{1}\bar{2}45, \quad 4:\bar{1}\bar{2}\bar{3}5, \quad 5:\bar{1}\bar{2}\bar{3}\bar{4}$$

of the CC system corresponding to a pentagon. The cocktail-shaker sort network defines the tournaments

$$1:2345, \quad 2:\bar{1}534, \quad 3:\bar{1}5\bar{2}4, \quad 4:\bar{1}5\bar{2}\bar{3}, \quad 5:\bar{1}4\bar{3}\bar{2};$$

this is isomorphic to the middle example of (5.1), with point 5 at the bottom and points $(2, 3, 4, 1)$ at the top. The odd-even transposition sort defines

$$1:2345, \quad 2:354\bar{1}, \quad 3:\bar{2}\bar{1}54, \quad 4:5\bar{2}\bar{1}\bar{3}, \quad 5:\bar{4}\bar{2}\bar{1}\bar{3},$$

another pentagon. The third example of (5.1) can be defined by the network

$$(8.3)$$

with tournaments

$$1:2345, \quad 2:\bar{1}453, \quad 3:\bar{1}45\bar{2}, \quad 4:\bar{1}\bar{3}\bar{2}5, \quad 5:\bar{1}\bar{3}\bar{2}4.$$

Notice that the third diagram in (8.1) shows pairs of transpositions above each other. When $|i - j| > 1$, the transpositions $[i, i + 1]$ and $[j, j + 1]$ commute; they can be performed in either order, or simultaneously, without changing their effect. Similarly, two ordered pairs $\frac{p}{q} \, \frac{r}{s}$ can be interchanged in a betweenness arrangement without affecting the corresponding CC system, if $\frac{p}{q}$ and $\frac{r}{s}$ have no points in common. We call two reflection networks *equivalent* if they can be obtained from each other by interchanging transpositions that commute.

One way to eliminate the effect of commutation is to bring each transposition as far to the left as possible. For example, (8.2) is converted in this way to the following compressed form:

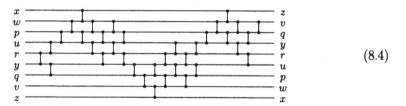

(8.4)

Equivalent networks have the same compressed form. The odd-even transposition sort has the shortest compressed form, among all reflection networks for n points; the cocktail-shaker sort has one of the longest, because it doesn't compress at all.

Another way to avoid the effect of commutativity is to insist that $[i, i + 1]$ be followed by $[j, j + 1]$ only if $j \leq i + 1$. (If $j > i + 1$, the transposition $[j, j + 1]$ should be done first, perhaps even before the predecessor of $[i, i + 1]$.) This gives a canonical order; two reflection networks whose transpositions appear in canonical order are equivalent if and only if they are identical. The canonical order corresponding to (8.4) is

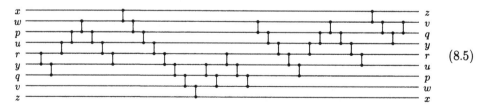

(8.5)

The CC system defined by an arrangement is unchanged if we remove the first pair $\frac{p}{q}$ and append $\frac{q}{p}$ as a new last pair. The corresponding operator on reflection networks removes the first transposition $[i, i + 1]$ and appends $[n - i, n - i + 1]$ at the end. Reflection networks are called *weakly equivalent* if they can be obtained from each other by commutativity and/or end-around moves. For example, the odd-even transposition sort is weakly equivalent to bubblesort in (8.1), because we can move four of its transpositions from the upper left to the lower right.

There is an almost-canonical form for reflection networks under weak equivalence, giving one canonical reflection network for each *extreme point* of the corresponding CC system. An extreme point is a point that is smallest or largest in the ordering induced by an arrangement; equivalently, it is a point that appears in the top or bottom line of the corresponding reflection network. It is not difficult to see that the number of extreme points is exactly the number of transpositions $[1,2]$ that occur between the top two rows plus the number of transpositions $[n-1,n]$ between the bottom two rows.

If x is an extreme point, we can move transpositions from front to back until we get to the transposition $[1,2]$ that moves x down from the top row. Let $[1,2]$ be the first transposition of our almost-canonical form. Then the reflection network will contain a first occurrence of $[2,3]$, which moves x down once more; there will be a first occurrence of $[3,4]$ after that, when x moves again; and so on. All transpositions that happen to be intermixed with those that move x are disjoint from those that move x, so we can commute them to the left and then remove them to the right. Therefore we can assume that the first $n-1$ transpositions are

$$[1,2]\,[2,3]\,\ldots\,[n-1,n]$$

in this order. The remaining $\binom{n}{2} - (n-1) = \binom{n-1}{2}$ transpositions now define a reflection network on $n-1$ elements. That network should be put into the canonical form described earlier, in which $[j, j+1]$ follows $[i, i+1]$ only if $j \le i+1$. This is the *almost-canonical form* for weak equivalence, promised above, given the extreme point x.

Incidentally, it is not difficult to prove that any reflection network in canonical form begins with $[1,2]\,[2,3]\,\ldots\,[n-1,n]$ whenever its first transposition is $[1,2]$. In fact, the first appearance of $[1,2]$ in any canonical reflection network is always followed immediately by $[2,3]\,\ldots\,[n-1,n]$.

Two reflection networks that begin with $[1,2]\,[2,3]\,\ldots\,[n-1,n]$ define the same CC system on $\{1, 2, \ldots, n\}$ if and only if the networks on $n-1$ defined by the remaining $\binom{n-1}{2}$ transpositions are (strongly) equivalent. Otherwise the networks would define different tournaments. Thus, two reflection networks can be tested for weak equivalence by putting one of them in almost-canonical form and seeing if that network is identical with any of the almost-canonical forms corresponding to extreme points of the other. If so, the two networks are weakly equivalent. If not, they're not.

For example, it turns out that all of the almost-canonical forms of bubblesort (the pentagon) are identical, namely

$$\text{(8.6)}$$

the cocktail-shaker sort on 5 elements has three almost-canonical forms,

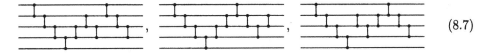

$$\text{(8.7)}$$

and (8.3) has four:

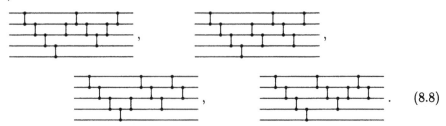

$$\text{(8.8)}$$

These are all of the almost-canonical forms on five points. And if we strike out the first four transpositions and the bottom line, we obtain canonical representations of all the (strong) equivalence classes on four points. The number of distinct almost-canonical forms corresponding to a given CC system is a divisor of the number of extreme points, because each almost-canonical form has a unique "successor" after making end-around moves.

These observations yield the following theorem:

Theorem. *Two arrangements of $\binom{n}{2}$ pairs satisfying the betweenness condition yield the same CC system if and only if one can be obtained from the other by interchanging disjoint pairs and/or removing the first pair $\frac{p}{q}$ and appending $\frac{q}{p}$ at the end. The number of CC systems on $\{1, 2, \ldots, n\}$ such that npq holds iff $n > p > q$ is the number of equivalence classes of reflection networks on $n - 1$ elements. The number of nonisomorphic CC systems on n points is the number of weak equivalence classes of reflection networks on n elements.*

Proof. If two arrangements define the same CC system, we can use the stated operations to transform the associated networks into almost-canonical form for some extreme point x.

These forms must be identical, or different tournaments will be defined. A CC system on $\{1, 2, \ldots, n\}$ with the stated property defines and is defined by a unique arrangement beginning with $\frac{n}{n-1} \ldots \frac{n}{2} \frac{n}{1}$, up to interchange of disjoint pairs. $\square$

The left-right mirror image of a reflection network is a reflection network that corresponds to an *anti-isomorphic* CC system: If pqr is true in the CC system defined by the original network, then pqr is false in the new system, and conversely. All reflection networks on 5 or fewer elements are weakly equivalent to their mirror images, but the following 6-element network is not:

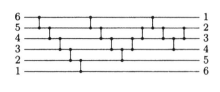
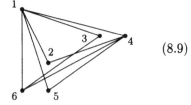

$$\text{(8.9)}$$

The corresponding CC system is realizable by a diagram such as the one shown; the mirror reflection of the diagram defines a CC system that is anti-isomorphic, but not isomorphic, to the unreflected system.

Reflection networks can be transformed into reflection networks in yet another way. When the first transpositions are $[1, 2] \ldots [n-1, n]$, as in an almost-canonical form, it is legitimate to reflect each of the others about a horizontal axis, thus replacing $[i, i+1]$ by $[n-i-1, n-i]$. Applying this to the bubblesort network, for example,

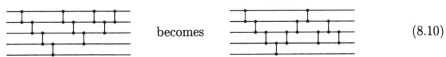

$$\text{becomes} \qquad\qquad\qquad\qquad\qquad (8.10)$$

Let us say that reflection networks are *preweakly equivalent* if they can be transformed into each other by using this flip operation together with the operations associated with weak equivalence. (The adjective "preweakly" is admittedly a somewhat weak contribution to mathematical terminology, but we will see in a moment that flipping is associated with preisomorphism.)

The flip operation has an easily understood affect on the corresponding CC systems. If we follow it by $\binom{n}{2}$ end-around shifts we obtain a network that is like the original unflipped network except that the first transpositions are changed from $[1, 2] \ldots [n-1, n]$ to $[n-1, n] \ldots [1, 2]$, and all other transpositions are shifted down one. Suppose the corresponding arrangement of pairs begins $\frac{n}{n-1} \ldots \frac{n}{2} \frac{n}{1}$ before shifting down; the corresponding arrangement after shifting is obtained by changing these to $\frac{1}{n} \frac{2}{n} \ldots \frac{n-1}{n}$. The betweenness condition still holds because we have $p > q$ in every other pair $\frac{p}{q}$. The effect on tournaments is obtained by changing n to $\bar{n}$ in the strings $\alpha_1, \ldots, \alpha_{n-1}$, and by replacing the string $\alpha_n = \overline{n-1} \ldots \overline{2}\,\overline{1}$ by $\overline{\alpha_n} = 1\,2 \ldots (n-1)$. This is precisely equivalent to what we would obtain by the preisomorphism that takes $(1, \ldots, n-1, n) \mapsto (1, \ldots, n-1, \bar{n})$; thus the CC systems are preisomorphic.

The discussion at the end of section 5 proves, in fact, that any two preisomorphic CC systems correspond to reflection networks that are preweakly equivalent.

Reflection networks have an important relationship to *simple arrangements of pseudolines* as defined by Levi in 1926 [52] (see Grünbaum's exposition [35]). Such an arrangement consists of n simple closed curves in the projective plane, with the property that every pair of curves intersects exactly once; the $\binom{n}{2}$ intersection points must also be distinct. Thinking of the projective plane as a sphere with antipodal points identified, we can stretch and bend each curve so that it stays roughly parallel to the equator, and so that all intersections occur in the "western hemisphere." Then a Mercator-like projection of these curves as seen on a map of the western hemisphere will look just like a reflection network, except that the transposition modules are changed into crossovers. For example, the 5-point bubblesort network of (8.6) is the same as the pseudoline arrangement

$$\begin{array}{l}
a \\ b \\ c \\ d \\ e
\end{array}
\qquad\qquad\qquad
\begin{array}{l}
e \\ d \\ c \\ b \\ a
\end{array}
\qquad , \qquad\qquad (8.11)$$

where a, b, c, d, e are antipodal points at the boundary of the hemisphere.

Interchanging crossovers that do not overlap preserves the arrangement; moving a crossover from the left to the right (and turning it upside down) preserves it too, if we rotate the lines slightly about the polar axis. The "flip" transformation also preserves a pseudoline arrangement; this corresponds to moving the line at upper left and lower right up and past the north/south pole. (The pseudolines divide the projective plane into $\binom{n}{2} + 1$ cells, one to the right of each crossover and one at the pole; the extreme points correspond to the lines that touch the cell containing the pole. Flipping, which is equivalent to shifting down, moves the pole into the cell that appears at the top left and bottom right in the unshifted network.)

An arrangement of pseudolines is called *stretchable* if the lines can all be made straight without changing the topological configuration of cells. Stretchable arrangements correspond to realizable CC systems; hence a CC system that is preisomorphic to a realizable CC system is realizable.

Corollary. *The number of nonpreisomorphic CC systems on n points is the number of preweak equivalence classes of reflection networks on n elements, and it is also the number of topologically different simple arrangements of n pseudolines in a projective plane.*

9. Enumeration

Let A_n be the total number of reflection networks on n elements, and let B_n, C_n, D_n, E_n be the corresponding number of equivalence classes, weak equivalence classes, weak equivalence/anti-equivalence classes, and preweak equivalence classes. Stanley [67] proved the remarkable theorem that

$$A_n = \frac{\binom{n}{2}!}{1^{n-1}\,3^{n-2}\,5^{n-3}\,\ldots\,(2n-3)^1}\,; \tag{9.1}$$

instructive combinatorial and algebraic explanations of this formula have been found by Edelman and Greene [16], Lascoux and Schützenberger [50]. Computer calculations give the following numerical results for small n:

n	1	2	3	4	5	6	7	8	9
A_n	1	1	2	16	768	292864	1100742656	48608795688960	29258366996258488320
B_n	1	1	2	8	62	908	24698	1232944	112018190
C_n	1	1	1	2	3	20	242	6405	316835
D_n	1	1	1	2	3	16	135	3315	158830
E_n	1	1	1	1	1	4	11	135	4382

We have seen in section 8 that C_n is the number of nonisomorphic CC systems on n points. In section 10 we will prove that D_n is the number of nonisomorphic uniform acyclic oriented matroids of rank 3 on n elements. This quantity D_n is also the number of topologically distinct, simple arrangements of pseudolines with a marked cell, as discussed by Goodman and Pollack [30].

The numbers B_n and C_n are related by

$$B_{n-1}/n \leq C_n \leq B_{n-1}, \tag{9.2}$$

because a weak equivalence class on n elements has at most n almost-canonical forms, and there are B_{n-1} almost-canonical forms. Obviously

$$C_n/2 \le D_n \le C_n. \tag{9.3}$$

We also have

$$D_n/\left(\binom{n}{2}+1\right) \le E_n \le D_n, \tag{9.4}$$

because we get at most $\binom{n}{2}+1$ preweakly inequivalent networks from a given weak equivalence/anti-equivalence class by moving the pole into each cell of the corresponding pseudoline arrangement. (This is much better than the obvious bound $C_n/2^n \le E_n$ that we get by simply counting the number of ways to negate points. Most point-negations give a pre-CC system that violates Axiom 4.)

The number A_n is asymptotically $2^{\Theta(n^2 \log n)}$. Indeed, $\log \binom{n}{2}! = \binom{n}{2} \log \binom{n}{2} + O(n^2) = n^2 \log n + O(n^2)$, and if n is even we have

$$\log \prod_{k=0}^{n-1} (2n-2k-1)^k = \sum_{k=0}^{n-1} k \log(2n-2k-1)$$

$$< \sum_{k=0}^{n/2-1} k \log 2n + \sum_{k=1}^{n/2} (n-k) \log n$$

$$< \sum_{k=1}^{n/2} n \log 2n = \tfrac{1}{2} n^2 \log 2n.$$

The table above indicates that B_n is substantially smaller than A_n, and indeed we can easily prove that

$$B_n < 2^{n^2+n}, \tag{9.5}$$

based on the canonical forms described in section 8. Every canonical form is a sequence of transpositions $[i_1, i_1+1] \ldots [i_l, i_l+1]$ where $l = \binom{n}{2}$ and $i_{k+1} \le i_k + 1$ for $1 \le k < l$; and there are fewer than $4^{l+n} = 2^{n^2+n}$ such sequences with $i_1 < n$, whether they correspond to reflection networks or not. For if we write down i_1 left parentheses, then for $1 \le k < l$ append $i_k - i_{k+1} + 1$ right parentheses and another left parenthesis, and finish with i_l right parentheses, we obtain a balanced string of $l + i_1 - 1$ matched parenthesis pairs from which $i_1 \ldots i_l$ can be reconstructed. The number of such strings with m matched pairs is the Catalan number $\frac{1}{m+1}\binom{2m}{m}$, which is less than 4^m.

Thus $\log B_n / \log A_n \to 0$ as $n \to \infty$. We have seen in (4.11) that weak pre-CC systems have the same asymptotic behavior as A_n, namely $2^{\Theta(n^2 \log n)}$; hence Axiom 1 has a strong effect on the total number of systems.

We can also prove a lower bound for B_n, having an asymptotic growth like (9.5) except for the coefficient of n^2 in the exponent. The following construction is based on the odd-even transposition sort (8.1); we need some notational conventions in order to describe it precisely. If α is a sequence of transpositions $[i_1, i_1+1] \ldots [i_m, i_m+1]$, let $\alpha + c$ denote the sequence $[i_1 + c, i_1 + c + 1] \ldots [i_m + c, i_m + c + 1]$. Suppose $a < b$, and let $\sigma_{a,b}$ be the sequence of $b - a$ transpositions $[a, a + 1][a + 1, a + 2] \ldots [b - 1, b]$. If $a < i < b$, we have $[i, i+1]\sigma_{a,b} = \sigma_{a,b}[i - 1, i]$; therefore if all transpositions $[i, i+1]$ in a sequence α satisfy $a < i < b$, we have

$$\alpha\sigma_{a,b} = \sigma_{a,b}(\alpha - 1). \tag{9.6}$$

If α is a sequence of m disjoint transpositions $[i, i+1]$, all with $a < i < b$, there are 2^m ways to write $\alpha = \alpha'\alpha''$; and for each of these we have $\alpha\sigma_{a,b} = \alpha'\sigma_{a,b}(\alpha'' - 1)$. Moreover, these 2^m subnetworks are inequivalent; this will be the key to constructing a large number of inequivalent reflection networks.

Lemma. $B^n \geq 2^{n^2/6 - O(n)}$.

Proof. Let $n = 2m$ where m is odd, and for $1 \leq k \leq m$ let $\alpha_k = [1, 2][3, 4] \ldots [m - 2, m - 1]$ if k is odd, $\alpha_k = [2, 3][4, 5] \ldots [m - 1, m]$ if k is even. We know that $\alpha_1\alpha_2 \ldots \alpha_m$ is a reflection network on m elements. It follows that

$$\sigma_{m,n}\sigma_{m-1,n-1} \cdots \sigma_{1,m+1}\,\alpha_1 \ldots \alpha_m\,(\beta + m)$$

is a reflection network on n elements, whenever β is a reflection network for m elements. By (9.6), this network can also be written

$$(\alpha_1 + m)\sigma_{m,n}(\alpha_2 + m - 1)\sigma_{m-1,n-1} \cdots (\alpha_m + 1)\sigma_{1,m+1}\,(\beta + m).$$

Furthermore, we can write each α_k as $\alpha'_k\alpha''_k$ in $2^{(m-1)/2}$ ways, giving $2^{m(m-1)/2} = 2^{n^2/8 - n/4}$ inequivalent reflection networks

$$(\alpha'_1 + m)\sigma_{m,n}(\alpha''_1 + m - 1)(\alpha'_2 + m - 1)\sigma_{m-1,n-1}(\alpha''_2 + m - 2) \ldots$$
$$(\alpha'_m + 1)\sigma_{1,m+1}(\alpha''_m + 1)(\beta + m). \tag{9.7}$$

Therefore $B_n \geq 2^{n^2/8 - n/4}B_{n/2}$, when $n \bmod 4 = 2$, and the lemma follows because we can prove by induction that $B_n \geq 2^{n^2/6 - 5n/2}$. $\quad\square$

Corollary. *The numbers B_n, C_n, D_n, and E_n all grow asymptotically as $2^{\Theta(n^2)}$.*

Proof. The bounds in (9.5) and the lemma establish this for B_n. Equations (9.2), (9.3), and (9.4) show that the other quantities are not substantially different. $\quad\square$

A somewhat sharper upper bound can be proved if we work harder. The main fact we need is that there aren't too many inequivalent networks in a weak equivalence class:

$$B_n \leq 3^n C_n. \tag{9.8}$$

Take a reflection network from some weak equivalence class, and append a copy of the same network but upside down; this gives a network of $n(n - 1)$ transpositions. Imagine that this network has been pasted on the equator of a sphere. The number of (strongly) inequivalent reflection networks that are weakly equivalent to the given one is at most the number of ways we can choose $\binom{n}{2}$ "consecutive transpositions" from this periodic double network. And this is the number of *cutpaths*, namely the number of southward paths from the north pole to the south pole, hitting each line once without touching any transposition modules. All transpositions between a cutpath and its antipodal mate form one of the networks enumerated by B_n.

The exact number of cutpaths can be computed by putting the number 1 in the cell at the north pole, then filling every other cell with the sum of the numbers in the adjacent cells lying to the north. The total number of cutpaths will then be the number in the cell at the south pole. For example, we get the following numbers from one of the networks constructed in the lemma above when $m = 3$ and $n = 6$:

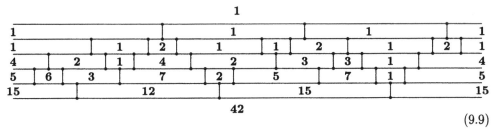

$$(9.9)$$

There are 42 cutpaths in this case.

The cells form a directed acyclic graph (a "dag") if we connect adjacent cells by southward-pointing arcs. The unique source vertex is the cell at the north pole; the unique sink vertex is the cell at the south pole; there are $n(n-1)$ other vertices. The cutpaths are the paths from source to sink. The arcs entering and leaving each vertex have a definite left-to-right order. (More precisely, the order is cyclic, but this distinction matters only at the source and the sink.)

Each arc of the dag can be labeled with a number from 1 to n, representing the name of the point currently occupying the line that is being crossed when we move from one cell down to another. (Equivalently, the arc label is the number of the corresponding pseudoline, if we regard the transposition modules as crossovers.) Each vertex can be labeled with the set of all arc numbers on the path from the source. (This is the set of all pseudolines crossed by that path.) The arc labels on every cutpath form a permutation of $\{1, 2, \ldots, n\}$, uniquely identifying the path.

We want to prove that there are at most 3^n cutpaths. This would certainly hold if all vertices were known to have outdegree ≤ 3, as in the example above; but some vertices might have high outdegree. We need to establish some special property of the dag if we are going to obtain a decent upper bound on the number of cutpaths. Otherwise we might, for example, have a dag with approximately n vertices on odd levels and just one vertex on every even level, in which case there would be approximately $n^{n/2}$ cutpaths.

The property we need depends on *middle arcs*, the arcs other than the leftmost or rightmost that lead out of a vertex whose outdegree is ≥ 3. All arcs from the source vertex can also be considered middle arcs. If p is the label of any middle arc leading from vertex v, we will prove that no path from v goes through any other middle arc labeled p.

Let $v \to u$ be a middle arc from v with label p. Suppose there is a path $v \to w \to^* x \to y$ where $x \to y$ is another middle arc labeled with p. Then $w \neq u$, because p cannot occur twice on a path. Notice that a middle arc always goes to a vertex of indegree 1 and outdegree ≥ 2. The transposition immediately to the left of cell y brings p up to the line above y, and the transposition immediately to the right of y brings p down again; thus, p zigzags when it labels a middle arc. (See, for example, the cells containing '1' in the second, third, and fourth rows of (9.9).)

Case 1, v is the source vertex. Then p is an extreme point, a point that moves from bottom to top to bottom in the reflection network, without zigzagging. Therefore p cannot be a middle label anywhere else. **Case 2**, w is to the left of u. Extend the path to a cutpath $n \to^* v \to w \to^* x \to y \to z \to^* s$, where n is the source, z is the rightmost child of y, and s is the sink. This cutpath defines a reflection network in which the first transposition moves p down and replaces it by q, the label on $y \to z$. Thereafter q stays above p. But q cannot be above p at the arc $v \to u$, because q would then appear as a label on the path $n \to^* v$. **Case 3**, w is to the right of u. This case is symmetrical to Case 2.

To complete the proof of (9.8), let $a_{m,r}$ be the maximum number of paths of length m from any vertex in such a dag when there are r permissible labels on middle arcs. Then $B_n \leq a_{n,n}C_n$, and we can show by induction on m that $a_{m,r} \leq 3^r 2^{m-r}$. The latter inequality is clear for $m = 0$, and when $m > 0$ we have

$$a_{m,r} \leq \max(1 \cdot a_{m-1,r}, 2 \cdot a_{m-1,r}, 3 \cdot a_{m-1,r-1}, 4 \cdot a_{m-1,r-2}, 5 \cdot a_{m-1,r-3}, \dots)$$

$$= \max(\tfrac{1}{2}, 1, 1, \tfrac{8}{9}, \tfrac{20}{27}, \dots) 3^r 2^{m-r} = 3^r 2^{m-r}.$$

(If the outdegree is $k + 2$, there will be k newly prohibited labels on middle arcs in the remaining paths of length $m - 1$.)

Theorem. *The number C_n of nonisomorphic CC systems is at most $3^{\binom{n}{2}}$.*

Proof. We have shown that $C_n \leq B_{n-1} \leq 3^{n-1}C_{n-1}$, and $C_1 = 1$. □

The quantity 3^n in (9.8) can probably be reduced to $n2^{n-2}$; at least, this is what we get from the bubblesort network (an n-gon), and the author has been unable to construct examples with a larger number of cutpaths. (It is *not* true that the number is bounded by 2^{n-2} times the number of extreme points.) Computer experiments for small n provide good support for the conjectured bound $n2^{n-2}$:

n	min	max	mean
4	12	16	14.0
5	22	40	31.3
6	36	96	58.9
7	56	224	106.5
8	82	512	194.5
9	116	1152	353.9

These mean values assume a uniform distribution over the equivalence classes enumerated by D_n. Of course, a sequence does not always reveal its true asymptotic behavior until the values become large.

Goodman and Pollack [31] have proved that the number of different *realizable* CC systems is only $2^{\Theta(n \log n)}$. Their upper bound depends on Milnor's theorem of algebraic geometry, which implies that the zeroes of a polynomial of degree d in k real variables always partition R^k into at most $(2 + d)(1 + d)^{k-1}$ connected components. Consider the polynomial in $(x_1, y_1, \ldots, x_n, y_n)$ that we obtain by multiplying $\binom{n}{3}$ distinct determinants $|pqr|$ together; this polynomial will vanish at the "boundaries" between nonisomorphic realizable CC systems. Hence the number of such systems is at most $\left(2 + 2\binom{n}{3}\right)\left(1 + 2\binom{n}{3}\right)^{2n-1}$.

Goodman and Pollack's lower bound argument is much more elementary. Given n points defining a realizable system, draw the $\binom{n}{2}$ lines connecting them, thereby obtaining $\frac{1}{3}n^3 - \frac{4}{3}n + 2$ cells. Put an $(n+1)$st point in "general position" in any of those cells, thus obtaining $\frac{1}{3}n^3 - \frac{4}{3}n + 2$ different systems on $n + 1$ labeled points. The total number of nonisomorphic realizable systems is therefore at least equal to $\prod_{k=2}^{n-1}\left(\frac{1}{3}k^4 - \frac{4}{3}k + 2\right)/n!$.

A similar lower bound follows, in fact, from a simple direct construction. Replace each point of an n-gon by a pair of points extremely close together. Rotating each pair of points independently gives at least $(n-1)^n/n$ nonisomorphic, realizable CC systems on $2n$ points.

10. Oriented matroids*

CC systems turn out to be equivalent to configurations that have arisen in yet another part of mathematics, where they are known as "uniform acyclic oriented matroids of rank 3." More precisely, there is a two-to-one correspondence between CC systems on a set of labeled points and all such oriented matroids defined on the same set. The two CC systems with the same image under this correspondence are obtained from each other by negating the value of every triple pqr.

The axioms for matroids are quite different from the axioms for CC systems, so it is worthwhile to study how the two kinds of systems are derivable from each other. First, we need some definitions. A uniform oriented matroid of rank 3 is a collection of *circuits*, which are sets of signed points $\{p, q, r, s\}$ with the following properties:

M1. If $\{p, q, r, s\}$ is a circuit, the absolute values $|p|$, $|q|$, $|r|$, $|s|$ are distinct.

M2. If $\{a, b, c, d\}$ is any set of four unsigned points, there is a circuit $\{p, q, r, s\}$ with $|p| = a$, $|q| = b$, $|r| = c$, $|s| = d$.

M3. If $C = \{p, q, r, s\}$ is a circuit, so is the negated set $\overline{C} = \{\bar{p}, \bar{q}, \bar{r}, \bar{s}\}$.

M4. If $C = \{p, q, r, s\}$ and $C' = \{\bar{p}, q', r', s'\}$ are any circuits with $C' \neq \overline{C}$, then there is at least one circuit C' contained in the set $\{q, r, s, q', r', s'\}$.

An oriented matroid is called *acyclic* if every circuit contains at least one negative point (hence at least one positive point, by M3).

The theory of oriented matroids is quite extensive, and it deals with considerably more general systems than these. (See [6], [21], and [48].) The general definition is like

---

* This section is independent of the rest of the monograph, except for sections 20 and 21, and it can be omitted on first reading.

the one above except that circuits are allowed to contain any number of signed points; then Axiom M2 is replaced by the statement that no circuit is properly contained in another. The theory is motivated by the study of linear dependence in a vector field over an order field: Equations of the form $a_1 x_1 + \cdots + a_r x_r = 0$ in which any $r - 1$ of the x_j are linearly independent define the sets $\{\operatorname{sign}(a_1)x_1, \ldots, \operatorname{sign}(a_r)x_r\}$, which are circuits of an oriented matroid. Our purpose here is to look closely at one small corner of the theory, which corresponds to CC systems. For brevity we'll call a collection of circuits satisfying M1–M4 a *4M system*.

Axioms M1, M2, M3 boil down to saying that every circuit is obtained by attaching signs to a four-point set, and that every four-point set of unsigned points corresponds in this way to at least two circuits (which are negatives of each other). In fact, every four-point set corresponds to *exactly* two circuits C and $\overline{C}$. For if we could have, say, $C = \{p, q, r, s\}$ and $C' = \{\bar{p}, q, r, \bar{s}\}$, then M4 would imply the existence of $C'' \subseteq \{q, r, s, \bar{s}\}$, which is impossible.

The connection between 4M systems $\mathcal{M}$ and CC systems $\mathcal{C}$ is quite simple: The circuits $\{p, q, r, s\}$ of $\mathcal{M}$ are precisely the sets of signed points such that the signed triples of $\mathcal{C}$ satisfy

$$sqp = srq = spr = pqr, \tag{10.1}$$

where $=$ denotes equality of boolean values (all true or all false).

Although rule (10.1) is simple, we have to show that it makes sense. In the first place we need to verify that the equations are symmetric in p, q, r, s: If we interchange any two variables, the two triples containing them are negated, and the other two triples change places while being negated; so equality is indeed preserved. Moreover, if we negate all four points, we negate all four triples; hence $\{p, q, r, s\}$ is a circuit if and only if $\{\bar{p}, \bar{q}, \bar{r}, \bar{s}\}$ is.

Finally, we can verify that we obtain exactly one system of triples satisfying Axioms 1–3 if we start with a 4M system and use (10.1) to define counterclockwise relations, assuming that a particular triple abc is true. (Another system, with abc false, will also be defined, having all triples complemented.) Suppose we have defined all triples consistently on a subset S containing at least three points. This is true initially with $S = \{a, b, c\}$. If $p \notin S$, we can define all triples on $S \cup \{p\}$ as follows: To define pqr, let q, r, s be any signed points of S, with signs chosen so that $\{p, q, r, s\}$ is a circuit. Then we take $pqr = srq$. This definition, which agrees with (10.1), does not depend on s. For if $\{p, q', r', t\}$ is another circuit, with $|q'| = |q|$ and $|r'| = |r|$ and $|t| \neq |s|$, then $\{\bar{p}, \bar{q}', \bar{r}', \bar{t}\}$ is also a circuit by M3, hence by M4 there is a circuit $C \subseteq \{q, \bar{q}', r, \bar{r}', s, \bar{t}\}$ Let $C = \{q'', r'', s, \bar{t}\}$, where $|q''| = |q|$ and $|r''| = |r|$. Then (10.1) implies that $sr''q'' = \bar{t}q''r'' = tr''q''$, and it follows that $srq = trq$. Thus pqr gets the same value via point t as it does via point s.

For example, let's consider the triples on four points $\{a, b, c, d\}$ that can be obtained from a 4M system. In this case there are just two circuits, C and $\overline{C}$. If $C = \{a, b, c, \bar{d}\}$, the triples according to (10.1) satisfy $dab = dbc = dca = abc$; so we have $d \in \Delta abc$ if those triples are all false, and $d \in \Delta cba$ if they are all true. Another case arises when $C = \{a, \bar{b}, c, \bar{d}\}$; then we have $dba = dca = dcb = acb$, which is a 4-gon with b opposite d and a opposite c; the vertices in clockwise order are either $abcd$

or *cbad* depending on whether the triples are true or false. Finally if $C = \{a, b, c, d\}$, we have triples $dba = dcb = dac = abc$ that violate Axiom 4. Therefore Axiom 4 holds if and only if the 4M system is acyclic as defined above.

We will show that any 4M system (possibly cyclic) defines a pre-CC system; thus, pre-CC systems are in two-to-one correspondence with systems of circuits that satisfy M1–M4. In fact it turns out that the full power of Axiom M4 is not needed. We will be able to deduce Axiom 5 from two very special cases of that axiom, once we have used it as above to establish a well-defined system of triples satisfying Axioms 1–3. The derivation begins with an intermediate result:

Axiom 6. $\neg(tsp \wedge tsq \wedge tsr \wedge tpq \wedge tqr \wedge trp \wedge sqp \wedge srq)$.

The first six triples of this axiom are the same as Axiom 5; they state that the tournament for t has a vortex out of s. The two additional triples say that the tournament for s has two arcs $q \to p$ and $r \to q$ that go "against the grain" of the cycle arcs $p \to q \to r$ in the tournament for t. Thus, Axiom 6 is apparently weaker than Axiom 5. We can derive it from M4 by assuming $tsp \wedge tsq \wedge tsr \wedge tpq \wedge tqr \wedge sqp \wedge srq$ and proving that trp must then be false. The triples $tsq \wedge tsr \wedge tqr \wedge srq$ are the same as $t\bar{q}s \wedge tr\bar{q} \wedge tsr \wedge s\bar{q}r$, so $\{s, \bar{q}, r, t\}$ is a circuit by (10.1). The triples $tsp \wedge tsq \wedge tpq \wedge sqp$ are the same as $t\bar{p}s \wedge tq\bar{p} \wedge tsq \wedge s\bar{p}q$, so $\{s, \bar{p}, q, t\}$ is also a circuit. Therefore by M4, $\{s, \bar{p}, r, t\}$ must be a circuit: We must have $t\bar{p}s = tr\bar{p} = tsr = s\bar{p}r$. In particular, since tsr is known to be true, trp must be false.

Axiom 6 is related to the law of interior transitivity, (2.4), which says that if $q \in \Delta tsr$ and $p \in \Delta tsq$ then $p \in \Delta tsr$. In these terms, Axiom 6 says, "if $q \in \Delta tsr$ (i.e., if $tsq \wedge tqr \wedge srq$) and tsr and $p \in \Delta tsq$ (i.e., $tsp \wedge tpq \wedge sqp$) then tpr"; here tpr is half of the conclusion $p \in \Delta tsr$, as in (2.4b), the other half being psr. The hypothesis tsr, which does not appear in (2.4), follows from $q \in \Delta tsr$ if we assume Axiom 4. However, we know from section 2 that Axiom 4 and (2.4b), i.e., Axioms 4 and 6, are not strong enough imply to imply Axiom 5. At least one more consequence of M4 is needed.

Notice, however, that M4 is valid for *signed* points, while we have used only unsigned points in our derivation of Axiom 6. Therefore Axiom 6 is actually true for all combinations of signed points p, q, r, s, t. There is one symmetry that takes Axiom 6 into itself (negate s, negate t, and interchange p with r); otherwise the signed permutations of p, q, r, s, t yield $32 \times 5!/2 = 960$ different axioms, all of which are valid on any 5-point system of triples derived from a 4M system.

We need not use all this flexibility. It suffices to consider a single additional axiom, obtained from Axiom 6 by negating s, p, q, and r; the two triples not involving t then change sign:

Axiom 6′. $\neg(tsp \wedge tsq \wedge tsr \wedge tpq \wedge tqr \wedge trp \wedge spq \wedge sqr)$.

This axiom states that an out-vortex in one tournament cannot coexist with two arcs that go "*with* the grain." Together with Axiom 6, it implies that no out-vortex can exist; two out of three arcs must go one way or the other. Thus, Axioms 6 and 6′, together with Axioms 1–3, imply Axiom 5.

We have proved that every 4M system defines two complementary pre-CC systems. There remains a possibility that Axioms M1–M4 might imply even more; they might define a restricted class of pre-CC systems, because our proof did not apply the full power of Axiom M4. Therefore we want to show conversely that any given pre-CC system defines a 4M system, if we use relation (10.1) to define circuits.

First we will show that the circuits defined by (10.1), given any pre-CC system, satisfy the following three laws introduced by Jon Folkman [21, Section 5]:

L1. If $\{p,q,r,s\}$ is a circuit, the absolute values $|p|$, $|q|$, $|r|$, $|s|$ are distinct.

L2. If $\{a,b,c,d\}$ is any set of four unsigned points, there are exactly two circuits $\{p,q,r,s\}$ with $|p| = a$, $|q| = b$, $|r| = c$, $|s| = d$, and these circuits are negatives of each other.

L3. If $\{p,q,r,s\}$ is a circuit and if t is a signed point with $|t| \notin \{|p|,|q|,|r|,|s|\}$, then there is a circuit $\subseteq \{p,q,r,s,t\}$ containing t.

Axiom L1 (which is the same as M1) obviously holds. Axiom L2 is satisfied because the triples (sqp, srq, spr, pqr) run through all 16 combinations of true and false as (s,p,q,r) run through all 16 combinations of plus and minus. Axiom L3 is satisfied because we know that all pre-CC systems on five elements are preisomorphic to a pentagon. The circuits for the pentagon (3.6) on $\{1,2,3,4,5\}$ form a 1–cycle

$$\{1\bar{2}3\bar{4}\} - \{1\bar{2}3\bar{5}\} - \{1\bar{2}4\bar{5}\} - \{1\bar{3}4\bar{5}\} - \{2\bar{3}4\bar{5}\}$$
$$\mid \qquad\qquad\qquad\qquad\qquad\qquad\qquad\qquad \mid \qquad\qquad (10.2)$$
$$\{\bar{2}3\bar{4}5\} - \{\bar{1}3\bar{4}5\} - \{\bar{1}2\bar{4}5\} - \{\bar{1}2\bar{3}5\} - \{\bar{1}2\bar{3}4\}$$

in which each circuit $\{p,q,r,s\}$ has two neighbors, one containing the remaining point t and the other containing $\bar{t}$, as required by L3.

There also is a more direct, low-level proof that L3 holds. If $\{p,q,r,s\}$ is a circuit, we can assume (possibly negating p,q,r,s), that we have $sqp \wedge srq \wedge spr \wedge pqr$. To avoid a vortex in the tournament for s, we must have at least one of stp, stq, str true, and at least one false; without loss of generality, we can assume that $stp = stq = srt = tpq$. To avoid a vortex between t and the cycle $q \to r \to s \to q$ in the tournament for p, we must then have $ptr = pqt$. And this makes $\{p,r,s,t\}$ a circuit if tpq is true, or $\{q,r,s,t\}$ a circuit if $tqp \wedge tqr$, or $\{p,q,r,t\}$ a circuit if $tqp \wedge trq$.

Call a system satisfying Axioms L1–L3 a 4L system. We can prove directly that a 4M system is a 4L system; only Axiom L3 needs to be verified. If $\{p,q,r,s\}$ and $\{p',q',r',t\}$ are circuits, with primed variables indicating plus or minus, then there is nothing to prove if $p' = p$, $q' = q$, $r' = r$. Otherwise suppose $p' = \bar{p}$, and apply Axiom M4 to $\{p,q,r,s\}$ and $\{\bar{p},q',r',t\}$, obtaining a circuit $\{q'',r'',s,t\}$. Again we're done if $q'' = q$ and $r'' = r$; otherwise suppose $q'' = \bar{q}$, and apply Axiom M4 to $\{p,q,r,s\}$ and $\{\bar{q},r'',s,t\}$ to get a circuit $\{r'',p,s,t\}$. If $r'' = r$, we are done, otherwise $\{\bar{r},p,s,t\}$ is one easy step from victory.

A 3L system is like a 4L system but with 3-element circuits instead of 4-element circuits. We can obtain a 3L system from a 4L system by fixing a point s and letting $\{p,q,r\}$ be a circuit iff $\{p,q,r,s\}$ or $\{p,q,r,\bar{s}\}$ is a circuit in a given 4L system. Axiom L3 is satisfied; namely, if $\{p,q,r\}$ is a circuit and t is a signed point with

$|t| \notin \{ |p|, |q|, |r| \}$, then either $\{p, q, t\}$ or $\{p, r, t\}$ or $\{q, r, t\}$ is a circuit. For we can assume that $\{p, q, r, s\}$ was a circuit, and there will be difficulty only if $\{p, q, r, t\}$ was also a circuit. But then either $\{q, r, t, \bar{s}\}$ or $\{p, r, t, \bar{s}\}$ or $\{p, q, t, \bar{s}\}$ was a circuit.

Now we want to prove that every 4L system is a 4M system; only Axiom M4 needs to be verified. First we prove a very special case: If $\{p, q, r, u\}$ and $\{\bar{p}, q, r, v\}$ are circuits in a 4L system, then $\{q, r, u, v\}$ is also a circuit. The analogous result is easily proved in a 3L system. For if $\{p, q, u\}$ and $\{\bar{p}, q, v\}$ are circuits, but $\{q, u, v\}$ is not, Axiom L3 applied to $\{p, q, u\}$ and v yields $\{p, u, v\}$ and the same axiom applied to $\{\bar{p}, q, v\}$ and u yields $\{\bar{p}, u, v\}$, a contradiction. The proof in a 4L system can now be obtained by reducing the problem to the 3L systems in which we fix q and r; we deduce the existence of circuits $\{q', r, u, v\}$ and $\{q, r', u, v\}$, where $q' = q$ or $\bar{q}$ and $r' = r$ or $\bar{r}$. The only viable possibility is $q' = q$ and $r' = r$.

Folkman completed his proof that Axioms L1–L3 imply Axioms M1–M4 by establishing a stronger result of independent interest.

Theorem. *Any two circuits $\{p, q, r, s\}$ and $\{p, t, u, v\}$ of a 4L system are connected by a path of circuits contained in $\{p, q, r, s, t, u, v\}$, where two circuits are considered to be adjacent if they have all but one element in common.*

Proof. The analogous result is easy in a 3L system: Given circuits $\{p, q, r\}$ and $\{p, u, v\}$, if they are not identical we can assume that $|v| \notin \{ |q|, |r| \}$ and $|r| \notin \{ |u|, |v| \}$. If they are not adjacent, there is a two-step path between them by Axiom L3, unless $\{v, q, r\}$ and $\{r, u, v\}$ are circuits. And in the latter case, three steps suffice.

In a 4L system, the proof is by induction on the number of distinct elements in $\{p, q, r, s, t, u, v\}$, the result being trivial if this number is ≤ 5. Otherwise we can assume that $|v| \notin \{ |q|, |r|, |s| \}$. Consider the 3L system obtained by fixing p; this system contains the circuits $\{q, r, s\}$ and $\{u, v, w\}$, and by Axiom L3 it also contains either $\{q, r, v\}$ or $\{q, s, v\}$ or $\{r, s, v\}$, say $\{q, r, v\}$. We can therefore find a path of length $l \leq 4$ from $\{q, r, s\}$ to $\{t, u, v\}$, and this path can be lifted to a path that passes through circuits $\{p_j, x_j, y_j, z_j\}$, where $\{p_0, x_0, y_0, z_0\} = \{p, q, r, s\}$, $\{p_l, x_l, y_l, z_l\} = \{p, t, u, v\}$, $p_j = p$ or $\bar{p}$, $\{x_j, y_j, z_j\} \subseteq \{q, r, s, t, u, v\}$, and $\{x_{j+1}, y_{i+1}, z_{i+1}\}$ is adjacent to $\{x_j, y_j, z_j\}$. If each $p_j = p$, we are done. Otherwise there is a least $j \geq 0$ with $p_{j+1} = \bar{p}$ and a greatest $k < l$ with $p_k = \bar{p}$. By the special case of Axiom M4 already verified, we know that $\{x_j, y_j, z_j\} \cup \{x_{j+1}, y_{j+1}, z_{j+1}\} = \{P, Q, R, S\}$ and $\{x_k, y_k, z_k\} \cup \{x_{k+1}, y_{k+1}, z_{k+1}\} = \{P, T, U, V\}$ are circuits; we are justified in assuming that they contain a common point P, because they are both contained in the set $\{q, r, s, t, u, v\}$. In fact, they contain at least two common points, and they can be connected by induction. $\square$

Corollary. *Every 4L system is a 4M system.*

Proof. Given $C = \{p, q, r, s\}$ and $C' = \{\bar{p}, t, u, v\}$, with $C' \neq \bar{C}$, we can assume that $|v| \notin \{ |p|, |q|, |r|, |s| \}$. Hence there is a circuit $C'' \subseteq \{p, q, r, s, v\}$ containing v, and a path from C'' to C' contained in $\{p, \bar{p}, q, r, s, t, u, v\}$. The first circuit on this path that doesn't contain p doesn't contain $\bar{p}$ either. $\square$

This completes the proof of equivalence between 4M systems and pre-CC systems, hence between acyclic 4M systems and CC systems.

Oriented matroids have orthogonal duals, which are defined by "cocircuits"; the cocircuits of a 4M system turn out to be the $\binom{n}{2}$ complementary pairs of sets of signed points p where $|p| \notin \{a, b\}$, with p and q having the same sign iff $abp = abq$. Although this connection between cocircuits and counterclockwise triples is simpler than (10.1), there does not appear to be a proof of equivalence between pre-CC systems and the duals of 4M systems that is any simpler than the proof given here.

11. Convex hulls

We define the *convex hull* of a CC system to be the set of all ordered pairs ts of distinct points such that tsp holds for all $p \notin \{s, t\}$.

If ts is in the convex hull, the tournament for t is transitive, by Axiom 5. The conventions of section 4 tell us that the tournament for t can be defined by a string $\alpha = p_1 \ldots p_{n-1}$ of positive points, with $p_1 = s$. It follows that $p_{n-1}t$ is also in the convex hull. But tp_j and $p_k t$ do not belong to the convex hull for any $j > 1$ or any $k < n - 1$.

This argument implies that every extreme point (i.e., every point with a transitive tournament) appears exactly twice among the ordered pairs of the convex hull, once as the first element and once as the second; those pairs must then consist of a number of directed cycles. In fact, there is always a unique cycle:

Lemma. *The convex hull of a CC system on $n \geq 2$ points consists of ordered pairs that form a cycle,*

$$t_1 t_2, \ t_2 t_3, \ \ldots, \ t_{m-1} t_m, \ t_m t_1, \qquad m \geq 2. \tag{11.1}$$

Proof. First we prove that there is at most one cycle. If t is an extreme point with tournament defined by $\alpha = p_1 \ldots p_{n-1}$, and if $t's'$ is any ordered pair of the convex hull with $t' \neq t$ and $s' \neq t$, we must have $t' = p_j$ and $s' = p_k$ for some $j < k$, because $t's't$ is the same as $tt's'$. Hence all such ordered pairs can be numbered $t_1 t_2, t_2 t_3, \ldots, t_{m-2} t_{m-1}$, with $t_1 = p_1$ and $t_{m-1} = p_{n-1}$. The remaining pairs of the convex hull are $p_{n-1}t$ and tp_1, so we have a cycle of the form (11.1) with $t_m = t$.

It remains to be shown that there is at least one cycle, i.e., that the convex hull is nonempty. If there are only two points $\{a, b\}$, the convex hull is $\{ab, ba\}$. If there are $n > 2$ points, let p be one of them, and let $\{t_1 t_2, \ldots, t_{m-1} t_m, t_m t_1\}$ be the convex hull of the remaining $n - 1$ points; this set is nonempty, by induction on n. Suppose the convex hull of all n points is empty. Then we must have $pt_{k+1}t_k$ for $1 \leq k < m$, and $pt_1 t_m$. Let k be maximum such that $pt_k t_1$ holds; then $k \geq 2$ and $k < m$, and we have $pt_1 t_{k+1}$. Hence $p \in \Delta t_{k+1} t_k t_1$, and Axiom 4 yields $t_{k+1} t_k t_1$, contradicting the assumption that $t_k t_{k+1} q$ holds for all $q \notin \{t_k, t_{k+1}, p\}$. □

The proof of the lemma uses both Axiom 4 and Axiom 5, and this is no accident. For if p, q, r, and t violate Axiom 4, they define a system of triples with no convex hull. And if p, q, r, s, t violate Axiom 5, they define a system with ts in the convex hull but not pt, qt, rt, or st. Therefore Axioms 4 and 5 are necessary and sufficient to obtain a ternary relation in which all subsets have a convex hull satisfying (11.1), assuming that Axioms 1–3 hold.

We can have $m = 2$ in (11.1) only when $n = 2$; hence we were justified in previous sections when we asserted that every CC system on three or more points contains at least three extreme points.

Incidentally, we noted in section 4 that vortex-free tournaments are not characterized by their score vectors. The same is true for pre-CC systems. But Goodman and Pollack [29] have shown that CC systems are nicer in this respect. If we know, for each pair of points pq, the number of paths r such that pqr holds, then we can reconstruct the entire CC system. This follows because the pairs with score 0 form the convex hull; and a point p on the hull has a transitive tournament, so we can call the other points $q_0, \ldots, q_{n-2}$, where pq_j has score j. Now $pq_j q_k$ holds iff $j > k$; so we compute the scores for the reduced CC system with p removed, and repeat the process.

If $\{t_1 t_2, \ldots, t_{m-1} t_m, t_m t_1\}$ is the convex hull of a CC system, we can prove that

$$t_i t_j t_k \text{ whenever } i < j < k. \tag{11.2}$$

This is true by the definition of convex hull when $j = i+1$ or $j = k-1$. Otherwise, we can assume by induction on $k - i$ that $t_{i+1} t_j t_k$ and $t_i t_j t_{k-1}$ are true; the tournament for t_i would then contain a vortex $t_{k-1} \to t_k \to t_j \to t_{k-1}$ out of t_{i+1} if we had $t_i t_k t_j$. Therefore the cyclic sequence of extreme points $(t_1, \ldots, t_m)$ forms an m-gon: The counterclockwise relation $t_i t_j t_k$ holds if and only if

$$i < j < k \quad \text{or} \quad j < k < i \quad \text{or} \quad k < i < j, \tag{11.3}$$

as in (3.6).

Theorem. *Suppose points $(t_1, \ldots, t_m)$ of a CC system form an m-gon, and let p be another point. If p lies outside the m-gon, say $pt_1 t_m$, then there exist indices $1 \leq j \leq l < m$ such that*

$$pt_k t_{k+1} \text{ if and only if } j \leq k \leq l. \tag{11.4}$$

On the other hand if $p \in \Delta t_i t_j t_k$ for some $i < j < k$, we have $pt_m t_1$ and $pt_k t_{k+1}$ for $1 \leq k < m$.

Proof. If $pt_1 t_m$, the ordered pairs ts with $t \neq p$ and $s \neq p$ in the convex hull of $\{p, t_1, \ldots, t_m\}$ are precisely the pairs $t_k t_{k+1}$ such that $pt_k t_{k+1}$ holds. Since the convex hull is a cycle, these pairs must be consecutive and the full convex hull must include also pt_j and $t_{l+1} p$, where j and l are defined by (11.3).

On the other hand if $p \in \Delta t_i t_j t_k$, the tournament for p includes the cycle $t_i \to t_j \to t_k \to t_i$, so p is not an extreme point. The convex hull of $\{p, t_1, \ldots, t_m\}$ must therefore be a cycle that doesn't involve p, and the only suitable cycle is $\{t_1 t_2, \ldots, t_{m-1} t_m, t_m t_1\}$. $\square$

These observations lead to an efficient incremental algorithm to find the convex hull of any CC system. Suppose the points are numbered $\{a_1, \ldots, a_N\}$ in any order, where $N \geq 2$. We will find the convex hull of $\{a_1, \ldots, a_n\}$ successively for $n = 2, 3, \ldots, N$; the current convex hull $(t_1, t_2, \ldots, t_m)$ will be represented by putting t_1 in a separate place and keeping the ordered list $(t_2, \ldots, t_m)$ in a binary search tree of some kind [45]. Initially $m = n = 2$, $t_1 = a_1$, and $t_2 = a_2$. If $n < N$, increase n by 1, set $p = a_n$, and update the convex hull as follows: **Case 1**, $pt_1 t_m$. Set $j = 1, 2, \ldots$, until $j = m-1$ or $pt_j t_{j+1}$. Then set $l = m-1, m-2, \ldots$, until $l = j$ or $pt_l t_{l+1}$. Delete $\{t_{l+2}, \ldots, t_m\}$ from the tree. If $j = 1$, insert p at the right of the tree (i.e., after t_{l+1}); otherwise delete $\{t_2, \ldots, t_{j-1}\}$ from the tree and replace t_1 by p. **Case 2**, $pt_m t_1$. Let t_k be the root of the tree. Then do a tree search as follows: If $k = m$ or $t_1 pt_k$, decrease k so that the new t_k is the left child of the old; otherwise increase k so that the new t_k is the right child of the old. This search terminates either when we want to decrease k and t_k has no left child, or when we want to increase k and t_k has no right child; in the latter case, increase k by 1. Then it follows that $t_1 pt_j$ holds iff $j \geq k$. (We have essentially placed p among $(t_2, \ldots, t_m)$ in the transitive tournament for t_1, knowing that p will appear before t_m.) If $k > 2$ and $pt_{k-1}t_k$, we have $p \in \Delta t_1 t_{k-1} t_k$ and the convex hull does not need to be updated. Otherwise we have discovered that $pt_k t_{k-1}$, hence p lies outside the m-gon $(t_1, \ldots, t_m)$. Set $j = k-1, k-2, \ldots$, until $j \leq 2$ or $pt_{j-1}t_j$. Set $l = k, k+1, \ldots$, until $l = m$ or $pt_l t_{l+1}$. Then delete $\{t_{j+1}, \ldots, t_{l-1}\}$ from the tree and insert p between t_j and t_l. (If $j = k-1$ and $l = k$, no deletion is made, and p is inserted in the place where the missing child occurred during the tree search.)

If the binary search tree is maintained as a balanced tree of some kind, the total running time of this algorithm will never exceed $O(N \log N)$, because the total number of tree operations amounts to at most N searches, N insertions, and N deletions. (Each point is inserted at most once and deleted at most once.) We cannot hope for a better worst-case time estimate than this, because the problem of finding the convex hull is well known to include the sorting problem as a special case: If real numbers $(x_1, \ldots, x_N)$ are given, the convex hull of the N points $a_k = (x_k, x_k^2)$ will be $\{a_{p(1)} a_{p(2)}, \ldots, a_{p(N-1)} a_{p(N)}, a_{p(N)} a_{p(1)}\}$ where $x_{p(1)} < x_{p(2)} < \cdots < x_{p(N)}$.

12. Another algorithm

Convex hulls can also be found by an incremental method that uses much simpler data structures. The current convex hull is maintained in a doubly linked circular list $(t_1, \ldots, t_m)$, and the counterclockwise tests are controlled by a binary branching structure from which no deletions need to be made. We will look closely at this alternative algorithm in the present section, because it will help to clarify a similar algorithm for Delaunay triangulation that appears in section 18 below.

The binary branching structure needed by this algorithm is essentially a dag with vertices of outdegree at most 2. We will consider it to be an array consisting of at most $4N - 6$ *nodes*; each node has two parts (p, α), where p points to an element of the CC system and α is a nonnegative integer.

The nodes appear in pairs (p_k, α_k) and (p_{k+1}, α_{k+1}), representing a *branch instruction*. The meaning of instruction k is, intuitively, "if $pp_k p_{k+1}$ then go to α_k, else go to α_{k+1}." More precise interpretations of the meaning will be given shortly.

Each element t of the current convex hull is represented by three fields $pred(t)$, $succ(t)$, and $inst(t)$. Here $pred(t)$ and $succ(t)$ point to the predecessor and successor of t in the convex hull, and $inst(t)$ is the address k of a node (p_k, α_k) such that $p_k = t$ and $\alpha_k = 1$. There will be exactly one such node in the branching structure for every element of the convex hull, and it will be one half of an instruction that means "if ptt' then go to 1 else go to α_{k+1}," where t' is the predecessor of t; the code value $\alpha_k = 1$ means that we will have to update the convex hull in the vicinity of t.

Initially $m = n = 2$, and our convex hull on two points (a_1, a_2) is represented by two nodes

$$(p_0, \alpha_0) = (a_1, 1), \quad (p_1, \alpha_1) = (a_2, 1) \tag{12.1}$$

in the branching structure, where

$$\begin{aligned} pred(a_1) &= succ(a_1) = a_2, \quad inst(a_1) = 0\,; \\ pred(a_2) &= succ(a_2) = a_1, \quad inst(a_2) = 1\,. \end{aligned} \tag{12.2}$$

There is a special variable l, initially 2, which represents the address of the first unused node. The algorithm now proceeds as follows, for $n = 3, 4, \ldots, N$:

Step H1.   [Consider a new point.]   Set $p \leftarrow a_n$ and $k \leftarrow 0$.

Step H2.   [Follow instructions.]   If $p_k p p_{k+1}$, increase k by 1.   Then if $\alpha_k = 0$, terminate the updating process; p is not in the convex hull.   If $\alpha_k = 1$, go to Step H3; p will be in the convex hull.   Otherwise (i.e., if $\alpha_k > 1$), set $k \leftarrow \alpha_k$ and repeat Step H2.

Step H3.   [Remove obsolete hull points.]   Set $t \leftarrow p_k$ and $s \leftarrow pred(t)$; also set $\alpha_k \leftarrow l$. Then perform the following two loops:

> **Step H3a.**   Set $q \leftarrow pred(s)$; while $q \neq t$ and psq, set $\alpha_{inst(s)} \leftarrow l$, $s \leftarrow q$, and $q \leftarrow pred(s)$.

> **Step H3b.**   Set $q \leftarrow succ(t)$; while $q \neq s$ and pqt, set $t \leftarrow q$, $\alpha_{inst(t)} \leftarrow l$, and $q \leftarrow succ(t)$.

Finally set $succ(s) \leftarrow p$, $pred(p) \leftarrow s$, $succ(p) \leftarrow t$, $pred(t) \leftarrow p$.

Step H4.   [Compile new instructions.]   Create the following four new nodes beginning at address l:

$$\begin{aligned} (p_l, \alpha_l) &= (p, 1), \quad &(p_{l+1}, \alpha_{l+1}) &= (s, l+2)\,; \\ (p_{l+2}, \alpha_{l+2}) &= (t, 1), \quad &(p_{l+3}, \alpha_{l+3}) &= (p, 0)\,. \end{aligned} \tag{12.3}$$

Then set $inst(p) \leftarrow l$, $inst(t) \leftarrow l+2$, and $l \leftarrow l+4$; the updating process is now complete. $\square$

For example, suppose we have $a_1a_2a_3$. Then when $n = 3$ we will get to Step H3 with $k = 0$ and we will get to Step H4 with $s = a_2$, $t = a_1$. The current instructions will become

$$
\begin{aligned}
(p_0, \alpha_0) &= (a_1, 2)\,, & (p_1, \alpha_1) &= (a_2, 1)\,; \\
(p_2, \alpha_2) &= (a_3, 1)\,, & (p_3, \alpha_3) &= (a_2, 4)\,; \\
(p_4, \alpha_4) &= (a_1, 1)\,, & (p_5, \alpha_5) &= (a_3, 0)\,;
\end{aligned}
\tag{12.4}
$$

the current convex hull cycle will be (a_1, a_2, a_3), with $inst(a_1) = 4$, $inst(a_2) = 1$, and $inst(a_3) = 2$; and l will be 6. The new instructions (12.4) have the following meaning, when we come through Step H2 with a new point p:

($k = 0$) If a_1pa_2, set $k \leftarrow 1$ and go to Step H3 with $p_k = a_2$; otherwise set $k \leftarrow 2$ and continue.

($k = 2$) If a_3pa_2, set $k \leftarrow 3$ and then set $k \leftarrow 4$ and continue; otherwise go to Step H3 with $p_k = a_3$.

($k = 4$) If a_1pa_3, do no updating; otherwise go to Step H3 with $p_k = a_1$.

In other words, the instructions can be paraphrased as follows: "If a_1pa_2, point p is outside the hull at a_2. Otherwise if a_2pa_3, point p is outside the hull at a_3. Otherwise if a_3pa_1, point p is outside the hull at a_1. Otherwise point p is inside the hull."

If on the other hand we have $a_2a_1a_3$, the situation will be essentially the same except that a_1 and a_2 will be interchanged. In that case we will have $(p_0, \alpha_0) = (a_1, 1)$, $(p_1, \alpha_1) = (a_2, 2)$; but the behavior of Step H2 is the same if nodes 0 and 1 are interchanged (or in general if nodes $2k$ and $2k + 1$ are interchanged).

To prove the correctness of the algorithm, we easily verify that the stated invariant conditions on $pred(t)$, $succ(t)$, and $inst(t)$ hold, and that we get to Step H3 only when the counterclockwise predicate $pred(p_k)\,p\,p_k$ is true. The theorem of section 11 then validates the updating of the hull that occurs in Step H3. Thus the algorithm will be correct if we can prove that p is not in the convex hull whenever Step H2 terminates with $\alpha_k = 0$.

For this purpose, we prove that any execution of Step H2 that leads to node number $k = l + 3$ in (12.3) must occur for a point p' such that $p' \in \Delta spt$. Such a point p' must follow a path in the branching structure that leads to $k = l$, afer which the algorithm determines that $pp's$ and $tp'p$ hold. To complete the proof, we need to demonstrate $sp't$.

The only way to reach $k = l$ is to come through one of the nodes whose α part is set to l in Step H3. Step H3 finds a sequence of consecutive nodes $t_0, t_1, t_2, \ldots, t_r, t_{r+1}$ in the current convex hull such that p lies inside the edges t_0t_1 and t_rt_{r+1} but outside the edges $t_1t_2, \ldots, t_{r-1}t_r$; in other words, we have

$$
pt_0t_1, \quad t_1pt_2, \quad \ldots, \quad t_{r-1}pt_r, \quad pt_rt_{r+1}\,.
$$

(Possibly $t_{r+1} = t_0$, or $t_r = t_0$ and $t_{r+1} = t_1$.) Step H3 deletes the points $t_2, \ldots, t_{r-1}$ from the convex hull and ends with $s = t_1$ and $t = t_r$. The nodes whose α part is set to l are precisely $inst(t_2), \ldots, inst(t_r)$. And Step H2 reaches $k = inst(t_j)$ only if it has determined that $t_{j-1}p't_j$ holds.

Therefore the validity of the algorithm boils down to proving the following result:

Lemma. *Let $p', p, t_1, \ldots, t_r$ be distinct points of a CC system such that $r \geq 2$ and $t_i t_j t_k$ holds for $1 \leq i < j < k \leq r$. If*

$$t_1 p t_2, \quad \ldots, \quad t_{r-1} p t_r, \quad p p' t, \quad t_r p' p,$$

and if $t_{j-1} p' t_j$ holds for some j, $1 < j \leq r$, then $t_1 p' t_r$.

Proof. The point p' lies outside the r-gon $(t_1, \ldots, t_r)$, so by the theorem in section 11 we can have $p' t_1 t_r$ only if $t_1 p' t_2$ or $t_r p' t_{r-1}$. Thus it suffices to consider the case $r = 3$. But when $r = 3$, the hypotheses $t_1 t_2 t_3 \wedge t_1 p t_2 \wedge t_2 p t_3 \wedge p p' t_1 \wedge t_3 p' p \wedge p' t_1 t_3$ imply $t_2 \in \Delta t_3 t_1 p$, hence $t_3 t_1 p$ by Axiom 4; hence $t_1 t_2 p'$ by Axiom 5, otherwise the tournament for t_1 would contain a vortex $t_2 \to t_3 \to p' \to t_2$ out of p; and $t_3 t_2 p'$ by Axiom 5', otherwise the tournament for t_3 would contain a vortex from $t_2 \to p' \to t_1 \to t_2$ into p. (The case $r = 3$ is, in fact, equivalent to rule (2.7).)   □

The running time of this algorithm can be $\Omega(N^2)$ in the worst case. For example, if the points $(a_1, a_2, \ldots, a_N)$ form an N-gon in the stated order, point a_n will be inserted into the hull only after verifying $a_1 a_2 a_n$, $a_2 a_3 a_n$, $a_1 a_3 a_n$, $a_3 a_4 a_n$, $a_1 a_4 a_n$, $\ldots$, $a_{n-2} a_{n-1} a_n$, $a_1 a_{n-1} a_n$. However, if the points $\{a_1, a_2, \ldots, a_N\}$ are accessed in random order, the average running time turns out to be quite fast, regardless of the underlying CC system.

Theorem. *The average number of counterclockwise tests made by the algorithm above is at most $3N \ln N + O(N)$, assuming that each of the $N!$ ways to assign subscripts to the points $\{a_1, \ldots, a_N\}$ is equally likely.*

Proof. Steps H1, H3, and H4 take only $O(N)$ units of time, because elements enter or leave the convex hull at most once each. Therefore Step H2 is the "inner loop," and we can prove the theorem by considering how often each of the instructions (12.3) is executed.

If l is an even number ≥ 2, let c_l be the number of times the test $p_l p' p_{l+1}$ is performed with the result false, and let c_{l+1} be the number of times it is performed with the result true. Let $C = c_5 + c_9 + c_{13} + c_{17} + \cdots$ be the number of times Step H2 terminates with no need to update the hull; clearly $C < N$.

Let E_j be the set of all pairs of points pq such that pqp' is true for exactly j other points p'. Let e_j be the number of elements of E_j, and let $e_{<j} = e_0 + e_1 + \cdots + e_{j-1}$. We will see that the average values of the numbers c_l can be bounded as a function of the numbers $e_{<j}$.

If tp belongs to E_j, the test $p_{l+2} p' p_{l+3} = tp'p$ of (12.3) will yield a false result only if p' is one of the j points that makes tpp' true. And it will be performed only if t and p occur earlier than all j of those points, with t occurring before p; otherwise tp would never be part of the current convex hull leading to the compilation of instructions (12.3). The probability that t and p appear before j other given points is $1/(j+2)(j+1)$. Therefore the sum $c_4 + c_8 + c_{12} + c_{16} + \cdots$ is at most

$$S = \sum_{j \geq 0} \frac{j e_j}{(j+1)(j+2)} = \sum_{j \geq 1} \frac{(j-2) e_{<j}}{j(j+1)(j+2)} < \sum_{j \geq 1} \frac{e_{<k}}{j(j+1)}. \tag{12.5}$$

Similarly, if ps belongs to E_j, the first test $p_l p' p_{l+1} = pp's$ of (12.3) will yield a false result only if p and s occur before all j points that make psp' true, with s occurring before p; we conclude that $c_2 + c_6 + x_{10} + c_{14} + \cdots$ is at most S. Furthermore we have $c_3 + c_7 + c_{11} + c_{15} + \cdots = (c_4 + c_5) + (c_8 + c_9) + \cdots \leq S + C$, because $c_{l+1} = c_{l+2} + c_{l+3}$ in (12.3). Therefore the total number of tests in Step H2 is at most $c_0 + c_1 + 3S + 2C = 3S + O(N)$.

Recall that the *score* of a vertex q in a tournament is the number of vertices r such that $q \to r$. Thus $pq \in E_j$ if and only if q has score j in the tournament associated with p. We will bound $e_{<j}$ by proving the following result:

Lemma. *A vortex-free tournament on $n \geq 2j$ vertices has at most j vertices with score $< j$.*

Proof. Suppose first that $n = 2j$, and let α be any string of signed points defining the tournament. If α has no negative points, the scores are $0, 1, \ldots, n-1$, and exactly j of these are $< j$. Otherwise if α begins with a negative point, say $\alpha = \bar{v}\beta$, we obtain the same tournament from the string βv, which has one less negative point. Otherwise α has the form $\beta u \bar{v} \gamma$ where u is positive and $\bar{v}$ is negative. Suppose u has score s and v has score t. Then $s + t = n - 1 = 2j - 1$, because $p \to u$ iff $v \to p$ for all $p \notin \{u, v\}$.

Replace α by the string $\beta \bar{v} u \gamma$; this reverses the direction of the arc $v \to u$ and leaves all other arcs unchanged. Therefore the new score of u is $s + 1$ and the new score of v is $t - 1$. The number of elements with score $< j$ has not changed, because u's score increases from $j - 1$ to j iff v's score decreases from j to $j - 1$. Repeating these operations will lead eventually to a string with nothing but positive entries; hence *exactly* j elements of every vortex-free tournament on $2j$ vertices have a score that is $< j$.

If $n > 2j$, suppose there are l vertices with score $< j$. Delete any vertex whose score is j or more; at least one such vertex must exist, otherwise the total score of all vertices would be at most $(j-1)n < \binom{n}{2}$. The new tournament has $l' \geq l$ vertices with score $< j$, and we know by induction on n that $l' \leq j$.   □

The lemma proves that $e_{<j} \leq jN$ when $N > 2j$, because $e_{<j}$ is a sum over N vortex-free tournaments with $N - 1$ vertices each. Therefore we can bound the quantity in (12.5):

$$S < \sum_{j \geq 1} \frac{e_{<j}}{j(j+1)} \leq \sum_{1 \leq j < N/2} \frac{N}{(j+1)} + \sum_{j \geq N/2} \frac{N^2}{j(j+1)} = N \ln N + O(N). \quad (12.6)$$

This completes the proof.   □

Our description of the algorithm does not explain how to output the convex hull after all N points have been processed. One way is to introduce a new variable r, which always points to a hull vertex; initially r points to a_1, say. Then in Step H3, set $r \leftarrow p$.

This algorithm is analogous to quicksort because it does almost no work in the inner loop besides making clockwise comparisons, and because its behavior depends on the cumulative effect of repeated partitioning of data that is in random order.

13. Comparison of algorithms

Section 11 presented an algorithm for convex hulls that we may call *treehull*; section 12 presented another that we may call *daghull*. Both algorithms find a convex hull of an arbitrary (possibly unrealizable) CC system. Treehull can be implemented to take $O(N \log N)$ time on any given N-point CC system; daghull has expected time $O(N \log N)$ on any given system that is input in random order.

It is interesting to compare both methods to a much simpler algorithm that we might call *hull insertion*. Hull insertion operates by keeping a doubly linked circular list of the points $(t_1, \ldots, t_m)$ currently in the convex hull, just as daghull does, but it uses no branching structure or anything else. Initially the circular list contains only the first two points, a_1 and a_2:

$$pred(a_1) = succ(a_1) = a_2 , \qquad pred(a_2) = succ(a_2) = a_1 ; \qquad (13.1)$$

and we have a pointer r to a_1. The algorithm now does the following simple computation, for $n = 3, 4, \ldots, N$:

Step I1. [Consider a new point.]   Set $p \leftarrow a_n$ and $s \leftarrow r$.

Step I2. [Go around the hull.]   Set $t \leftarrow succ(s)$. If spt, go to Step I3. Otherwise, if $t = r$, no updating needs to be done since p lies inside the m-gon $(t_1, \ldots, t_m)$. Otherwise set $s \leftarrow t$ and repeat Step I2.

Step I3. [Remove obsolete hull points.]   If $s \neq r$, skip to Step I3b. Otherwise perform both of the following loops:

> **Step I3a.**   Set $q \leftarrow pred(s)$; while $q \neq t$ and psq, set $s \leftarrow q$ and $q \leftarrow pred(s)$. Then set $r \leftarrow q$.

> **Step I3b.**   Set $q \leftarrow succ(t)$; while $t \neq r$ and pqt, set $t \leftarrow q$ and $q \leftarrow succ(t)$.

Finally set $succ(s) \leftarrow p$, $pred(p) \leftarrow s$, $succ(p) \leftarrow t$, $pred(t) \leftarrow p$.   □

Thus, hull insertion tries each edge st of the current hull; if the new point lies outside some edge, it is inserted into the hull as a new extreme point, and it may cause adjacent points to be removed from the list. But if we get all the way around without finding the new point outside, the new point must lie inside.

Hull insertion is clearly inefficient when the convex hull is large. But the method is extremely easy to program, and it has very little "data structure overhead" per operation, so it is actually the method of choice when the number of points is small.

We will compare the algorithms by considering three things: the length of the corresponding programs (the number of statements to implement them in the programming language C, not counting implementation of the predicate pqr); the number of times they test a triple pqr to see if it has counterclockwise orientation; and the number of "mems" the programs use to access and update data structures. A *mem* is a memory reference: Whenever a field of a record is examined or changed, we add one mem to the running time. Thus, when a field value is fetched into a "register," one mem is changed; henceforth that value can be looked at again without further cost, until the register changes, because computations among a fixed finite number of registers do not reference memory. The author has found that mem units provide meaningful machine-independent comparisons between algorithms for numerous applications; furthermore, there is a simple way to instrument a C program so that it will count mems properly. (See [47].)

Three implementations of treehull were tested. All three use a doubly linked list structure as well as a tree to represent the current hull, because many operations of the algorithm depend on the successors or predecessors of extreme points. The simplest implementation uses ordinary binary search trees without any attempt at balancing; the second uses Aragon and Seidel's *treap* structure [2], which guarantees that the binary tree will have the probability distribution of random binary search trees after every operation; the third uses Sleator and Tarjan's *splay trees* [66], which guarantee $O(N \log N)$ total time in the worst case. The second implementation will be called "treaphull," and the third will be called "splayhull."

Four kinds of data were used in the author's tests. First was some realistic data, taken from the latitude and longitude coordinates of 128 cities in the United States and Canada; this data comes from the Stanford GraphBase [47]. The convex hull in this case turned out to involve 13 of the 128 cities (Vancouver, BC; Salem, OR; Santa Rosa, CA; San Francisco, CA; Salinas, CA; Santa Barbara, CA; San Diego, CA; Victoria, TX; West Palm Beach, FL; Worcester, MA; Saint Johnsbury, VT; Winnepeg, Manitoba; and Regina, Saskatchewan).

The second kind of data was uniformly distributed inside a square. The third kind was simply an n-gon, with points entering in random order. The convex hull was, of course, much larger in this case than it was in the others.

There was also a need to test the algorithms on data that was intentionally *nonrandom*: When points are considered in random order, incremental hull algorithms almost never need to decrease the size of the current convex hull, because a new point rarely displaces more than one former point. Therefore large portions of the code were never executed when convex hulls were found from the first three kinds of data. In the fourth data set the points came in 10 nested groups, with $N/10$ randomly chosen points per group. The points in the kth group were placed approximately on the edges of a square, with the side of the square proportional to k^2. More precisely, each point was either (a, b) or (b, a), chosen with 50/50 probability, where a was a uniform random integer between 0 and $100k^2$, and b was uniform either between 0 and 100 or between $100k^2 - 100$ and $100k^2$. Points of a new group would tend to wash out several points of an old group, therefore allowing all parts of each algorithm to be exercised.

The results are summarized in the table on the next page. Numbers in parentheses represent lines of code (i.e., statements in C); thus, hull insertion needs only 27 program statements, compared to 150 for splayhull. Each of the running time figures stands for a single run, not an average of many runs, so we should look more at relative magnitudes than at the precise numbers. The data for each job was the same for each algorithm, and of course each algorithm gave the same answer. The "ccs" represent calls on the counterclockwise predicate. Each cc test requires six memory

data	hull insertion (27) mems+ccs	daghull (55) mems+ccs	treehull (107) mems+ccs	treaphull (148) mems+ccs	splayhull (150) mems+ccs	hull size
128 cities	1390+1246	3543+1034	2451+754	2856+882	3440+1056	13
100 uniform	912+816	2028+579	1709+525	1869+579	1990+552	10
1000 uniform	13128+12916	21078+6811	17654+6211	18576+6573	19430+6929	16
10000 uniform	230288+229836	210795+69806	196109+75278	194857+72973	205128+77732	28
100 n-gon	2922+2523	4530+1079	3570+984	5311+1029	10713+1015	100
1000 n-gon	251630+247630	63204+16737	45052+14299	63939+15085	175928+15503	1000
10000 n-gon	25365670+25325671	859311+243106	552193+193636	718351+192442	2436850+206162	10000
100 nested	900+561	7544+2174	2427+583	2869+605	3571+604	9
1000 nested	12910+11435	151002+48855	22670+6847	25983+7248	31814+6925	16
10000 nested	192834+189667	2318715+769733	191261+68149	211047+74144	229121+75223	19

references to fetch coordinates, besides the arithmetic instructions needed to evaluate a determinant; these six memory references are not included in the count of mems.

As expected, hull insertion is intolerably poor on n-gons, but otherwise it is quite reasonable on small problems, and its easily written program is only $1/4$ the size of the program for treehull. Notice that hull insertion was in fact the fastest of all methods tested, when presented with 100 random points nested in groups of 10.

The daghull algorithm blows up on nonrandom nested groups, for obvious reasons; otherwise it performs pretty well, considering that it requires only half as much code as treehull. But its data-structure overhead cost, measured in units of mems per cc, does not turn out to be as low as we might have expected.

In these experiments treehull usually wins on speed, although the extension to treaps does not make the program a great deal more complex nor does it lead to excessive overhead for rebalancing. Perhaps there will be a way to prove that treehull without rebalancing always has an expected running time as good as that of treaphull; until then, some people may well prefer to use treaphull, for which we have rigorous time estimates in spite of the fact that it does not perform quite as well on the data considered here. Treaphull beat treehull on 10000 uniformly random points in the test reported above, but further experiments show that there actually is a good deal of variation in different runs. For example, tests were made using exactly the same 1000 points in the same order, changing only the source of random numbers used for priorities in the treap; this gave the following numbers of mems plus ccs in five different runs of treaphull:

$$194857+72973; \quad 235634+93539; \quad 184895+68389; \quad 206333+78960; \quad 190092+66886.$$

The tabulated results are consistent with the conjecture that random insertions and deletions on binary search trees tend to produce trees whose total path length is less than that of a purely random tree. That conjecture may be impossibly difficult to prove, although it is known to hold in a very special case of the trees containing only three elements [43].

In this application the extra program and overhead cost of splay trees is not compensated by improved performance. Other types of worst-case-balanced trees are more difficult to program and will presumably fare no better; it appears that treaps are the balancing method of choice when rigorous balance estimates are important.

14. Degeneracy

We began with the assumption that no three points are collinear. But we could have defined a more general kind of system in which each triple pqr has three values 'true, false, none' or '$+1, -1, 0$', where the third value means that $\{p, q, r\}$ do not define a triangle. Such a theory would turn out to be equivalent to the general study of acyclic oriented matroids with rank ≤ 3. But it is much easier to develop algorithms for convex hulls based on CC systems as we have defined them, because special cases need not then be considered. If collinear points are undesirable in the output, a simple postprocessing algorithm will remove them.

Therefore let us try to find a way to define the ternary relation satisfying Axioms 1–5, given an arbitrary set of points in the plane. We will assume that the points are presented in some definite linear order; our definition of pqr will depend on the order of the points as well as on their Cartesian coordinates (x_p, y_p), (x_q, y_q), (x_r, y_r). We will write $p \prec q$ if point precedes point q in the assumed linear order.

It is easy to satisfy Axioms 1–3. The values of pqr can be defined arbitrarily when $p \prec q \prec r$, and this implies the values in all other cases.

It is also fairly easy to satisfy Axiom 4, using a determinant-like operation. Let $f(p, q)$ be any real-valued function of the points p, q, and define pqr to be true if the quantity

$$\Delta(p, q, r) = f(p, q) + f(q, r) + f(r, p) - f(q, p) - f(r, q) - f(p, r) \qquad (14.1)$$

is greater than some nonnegative threshold value θ, or if $|\Delta| \leq \theta$ and $(p \prec q \prec r) \vee (q \prec r \prec p) \vee (r \prec p \prec q)$. We get the standard definition for noncollinear points in the plane if $f(p, q) = x_p y_q$ and $\theta = 0$, because $\Delta(p, q, r)$ is then the nonzero determinant $|pqr|$ of section 1. We get approximations to this definition, if $f(p, q)$ is an approximation to $x_p y_q$ and if θ is some error tolerance. The values added and subtracted in (14.1) are supposed to be computed without error. Axioms 1–3 are easily verified, and Axiom 4 is a consequence of the identities

$$\Delta(t, p, q) + \Delta(t, q, r) + \Delta(t, r, p) + \Delta(r, q, p) = 0. \qquad (14.2)$$

For if we have $tpq \wedge tqr \wedge trp \wedge rqp$ in violation of Axiom 4, we must have at most one $\Delta > \theta$; otherwise the Δ's could not sum to zero. Thus we may assume by symmetry that any counterexample must satisfy $\big((t \prec p \prec q) \vee (p \prec q \prec t) \vee (q \prec t \prec p)\big) \wedge \big((t \prec q \prec r) \vee (q \prec r \prec t) \vee (r \prec t \prec q)\big) \wedge \big((t \prec r \prec p) \vee (r \prec p \prec t) \vee (p \prec t \prec r)\big)$. But these inequalities are unsatisfiable by any linear ordering, so Axiom 4 must hold.

Axiom 5 is, unfortunately, much more delicate and difficult to guarantee. Approximations to the determinant, computed with either fixed-point or floating-point arithmetic, are almost certainly untrustworthy. But if we can compute determinants with perfect accuracy, there are some pleasant ways to obtain a bona fide CC system for any set of points in the plane.

Suppose first that the points are distinct. Then the definition of pqr above will satisfy Axioms 1–5 if we take $f(p, q) = x_p y_q$ and $\theta = 0$. To verify this, we start with the determinant identity

$$|tpq|\,|tsr| + |tqr|\,|tsp| + |trp|\,|tsq| = 0. \qquad (14.3)$$

The Δ function (14.1) leads to a similar identity, but the right-hand side in general is a sum of thirty "commutator terms" of the form $f(p, r)f(q, s) - f(p, s)f(q, r)$; when $f(p, q) = x_p y_q$, these commutators all reduce to $x_p y_r \cdot x_q y_s - x_p y_s \cdot y_q y_r = 0$, assuming exact arithmetic.

If Axiom 5 fails, with $tsp \wedge tsq \wedge tsr \wedge tpq \wedge tqr \wedge trp$, we consider cases based on the number of determinants $|tps|$, $|tqr|$, $|trp|$ that are positive. **Case 0**, $|tpq| =$

$|tqr| = |trp| = 0$. This case is ruled out by the argument we used to establish Axiom 4.
Case 1, $|tpq| = |tqr| = 0 < |trp|$. Point t must lie on the lines qp and qr, so this case
is impossible when the points are distinct. **Case 2**, $|tpq| > 0$ and $|tqr| > 0$. Identity
(14.3) implies that $|tsr| = |tsp| = 0$, so point s must lie on the lines tr and tp; another
impossibility. Axiom 5 cannot fail, when the points are distinct.

In practice it is desirable to allow repeated points as well as collinear points;
then algorithms will work without exception. A study of repeated points also gives
insight into the limiting cases that arise when approximate arithmetic breaks down.
However, *there is no general way to satisfy Axioms 1–5 with the definition above in
the presence of multiple points*, regardless of the linear ordering chosen. For example,
assume that points p, q, r form a counterclockwise triangle, and assume that there are
five identical points inside that triangle. At least two of those five points, say $t \prec s$,
must be consecutive in the assumed linear ordering. But then we have $tsp \wedge tsq \wedge$
$tsr \wedge tpq \wedge tqr \wedge trp$, contradicting Axiom 5.

There is, however, a simple and somewhat surprising way to define a CC system on arbitrary points in the plane. First we assume for convenience that $\prec$ is a
refinement of the lexicographic order relation $<$:

$$p < q \iff x_p < x_q \ \vee \ (x_p = x_q \wedge y_p < y_q); \qquad (14.4)$$

$$p < q \implies p \prec q \implies p < q \ \vee \ p = q. \qquad (14.5)$$

Then we define pqr to be true if and only if

$$|pqr| > 0 \ \vee \ \bigl(|pqr| = 0 \ \wedge \ \bigl(\Psi(p,q,r) \vee \Psi(q,r,p) \vee \Psi(r,p,q)\bigr)\bigr), \qquad (14.6)$$

where

$$\Psi(p,q,r) = \bigl((p \prec q \prec r) \wedge (p \neq q)\bigr) \ \vee \ \bigl((q \prec p \prec r) \wedge (p = q)\bigr). \qquad (14.7)$$

Here $p = q$ means $(x_p, y_p) = (x_q, y_q)$; we will use different variable names to denote
different points of the system, although different points may happen to be equal,
coordinate-wise.

Definition (14.6) takes the construction that worked for collinear but distinct
points and skews it slightly so that it works also for repeated points. Another way to
state the effect of (14.6) is, "The counterclockwise predicate pqr holds with respect
to a repeated point $p = q$ and a different point r iff $q \prec p \prec r$ or $r \prec p \prec q$." The
other case, $(q \prec r \prec p) \wedge (q \neq r)$ in $\Psi(q,r,p)$, is incompatible with the $\prec$ relation
because of (14.5). The points "to the left of the line pq" when $p = q$ and $p \prec q$, i.e.,
the points r such that pqr holds, are therefore

$$\{\, r \mid (r \prec p \wedge r \neq p) \vee p \prec r \prec q \,\}; \qquad (14.8)$$

the points "to the right" are the remaining ones,

$$\{\, r \mid (r \prec p \wedge r = p) \vee q \prec r \,\}. \qquad (14.9)$$

Notice that the definition is not symmetric between left and right.

Axioms 1–3 hold, because (14.6) has cyclic symmetry and because $\Psi(p, q, r) = \neg\Psi(q, p, r)$. Axiom 4 also holds; it could fail only if $|spq| = |sqr| = |srp| = |rqp| = 0$, because of (14.2). If all four determinants vanish, we can assume by symmetry that $p \prec q \prec r \prec s$. A simple case analysis now verifies Axiom 4: If $p \neq q \neq r$ we have $\square pqrs$; if $p \neq q = r$ we have $r \in \Delta pqs$; if $p = q \neq r$ we have $\square prsq$; and if $p = q = r$ we have $\square rqps$. (Recall that $r \in \Delta pqs$ means $rpq \wedge rqs \wedge rsp$; $\square pqrs$ means $pqr \wedge qrs \wedge rsp \wedge spq$.)

Lexicographic order is convenient because a convex combination of points lies lexicographically between them:

$$p < q \quad \text{and} \quad 0 < \alpha < 1 \implies p < \alpha p + (1 - \alpha)q < q. \tag{14.10}$$

We can use this fact to deduce certain relations between the $\prec$ ordering and the predicate pqr.

Lemma. *The counterclockwise triples defined in the plane by (14.6) contain no four-point configurations of the following kinds:*

$$p \prec q \prec r \prec s \wedge s \in \Delta pqr; \tag{14.11}$$

$$s \prec p \prec q \prec r \wedge s \in \Delta pqr; \tag{14.12}$$

$$p \prec r \wedge p \prec s \wedge q \prec r \wedge q \prec s \wedge \square prqs; \tag{14.13}$$

$$p \prec r \wedge q \prec r \wedge r \prec s \wedge rqs \wedge rsp \wedge rqp; \tag{14.14}$$

$$p \prec q \wedge q \prec r \wedge q \prec s \wedge qsp \wedge qpr \wedge qsr. \tag{14.15}$$

Proof. If sqp, relation (14.14) is a special case of (14.11), because $s \in \Delta rqp$; otherwise qsp and (14.14) reduces to (14.13). Similarly, (14.15) is a consequence of (14.12) or (14.13) according as psr is true or false. So we need only show that (14.11), (14.12), and (14.13) are impossible. The case analysis above rules them out whenever all four determinants $|spq|$, $|sqr|$, $|srp|$, $|pqr|$ are zero; in particular, they cannot occur if any two of $\{p, q, r, s\}$ are equal.

If $s \in \Delta pqr$ and $|pqr| = 0$, we have $|spq| = |sqr| = |srp| = 0$, because the determinants are nonnegative and sum to $|pqr|$. Thus we may assume that $|pqr| > 0$. But then s is a convex combination,

$$s = \frac{|sqr|}{|pqr|} p + \frac{|psr|}{|pqr|} q + \frac{|pqs|}{|pqr|} r,$$

which is incompatible with (14.11) or (14.12) unless $s = r$ or $s = p$.

Finally, assume $\square prqs$. Identity (14.2) yields

$$|prq| + |qsp| = |rqs| + |spr|,$$

hence we cannot have $|prq| = |qsp| = 0$ unless $|rqs| = |spr|$. Suppose $|prq| > 0$; then

$$s + \frac{|qsp|}{|prq|} r = \frac{|rqs|}{|prq|} p + \frac{|spr|}{|prq|} q,$$

and the pairs of coefficients on each side are nonnegative and have the same sum. Dividing by this common sum and applying (14.10) shows that (14.13) cannot hold unless either r or s equals either p or q. A similar argument applies when we have $|qsp| > 0$. $\square$

Theorem. *The ternary relation defined in (14.6) yields a CC system on an arbitrary multiset of points in the plane.*

Proof. We will give two proofs, one combinatorial/geometric and the other algebraic/analytic, because both proofs shed some light on the structure. Only Axiom 5 remains to be verified, so we shall assume $tsp \wedge tsq \wedge tsr \wedge tpq \wedge tqr \wedge trp$ and try to obtain a contradiction. We know that Axiom 5 holds whenever the points are distinct, so our first proof simply rules out all counterexamples when repeated points appear.

Axiom 4 tells us that pqr holds. Suppose $|pqr| = 0$. Then the four determinants on $\{p, q, r, t\}$ vanish, and our case analysis above shows that the only possibility with $t \in \Delta pqr$ is $p \prec q \prec t \prec r$, with $p \neq q = t$. If also $t = s$, we need $t \prec s$ to satisfy pts, but $s \prec t$ to satisfy qts. Hence $t \neq s$ but then qts implies $s \prec q$, contradicting rts. Therefore $|pqr| > 0$, and in particular the points $\{p, q, r\}$ must be distinct.

Suppose $t = s$. If $t \prec s$, the relations tsp, tsq, tsr imply that $p \prec t$, $q \prec t$, $r \prec t$ unless t equals p, q, or r. But that contradicts (14.11), since $t \in \Delta pqr$. Similarly, if $s \prec t$ we cannot have $t \prec p \wedge t \prec q \wedge t \prec r$. Therefore we may assume that $t = p$; but then we have either $r \prec t \prec p \prec s$ or $p \prec s \prec t \prec r$, both of which contradict trp. Thus $t \neq s$.

Suppose $t = q$; this is the most interesting case. The relations tpq and tqr imply that we have either $p \prec q \prec t \prec r$ or $r \prec t \prec q \prec p$. Now tsq gives either $s \prec q \prec t$ or $t \prec q \prec s$. But $p \prec t \wedge s \prec t \wedge t \prec r \wedge tsr \wedge trp \wedge tsp$ is impossible by (14.14); and $r \prec t \wedge t \prec p \wedge t \prec s \wedge tsr \wedge trp \wedge tsp$ is impossible by (14.15). Thus t must be different from p, q, r, and s.

The only remaining possibility is $s = q$. Then $|tsp| = |tqp| \leq 0$, so $|tsp| = |tpq| = 0$. But if $s \prec q$, the hypotheses tsq and tpq and tsp force $t \prec s \prec p \prec q$, contradicting $p \neq q$. Hence $s \succ q$; but then tsq and tsp force $q \prec s \prec p \prec t$, contradicting tpq. The first proof of Axiom 5 is complete.

The second proof is completely different; it is based on the idea of perturbation, which is well known in the theory of linear programming. (See [17] for a general treatment of perturbation, called "simulation of simplicity.") If $p_1 \prec p_2 \prec \cdots \prec p_n$ is any refinement of lexicographic order, replace the coordinates (x_k, y_k) of p_k by

$$p'_k = \left(x_k - \epsilon^{3n^2 - nk}, y_k + \epsilon^{3n^2 - (n+1)k} \right). \tag{14.16}$$

If $p \prec q \prec r$, the coordinates of points p', q', r' will be $(x_p - \delta_p, y_p + \epsilon_p)$, $(x_q - \delta_q, y_q + \epsilon_q)$, $(x_r - \delta_r, y_r + \epsilon_r)$, where we have

$$\epsilon_r > \epsilon_q > \epsilon_p > \epsilon_r^2 ; \tag{14.17}$$

$$\epsilon_r \delta_q > \epsilon_q \delta_r , \qquad \epsilon_r \delta_p > \epsilon_p \delta_r , \qquad \epsilon_q \delta_p > \epsilon_p \delta_q ; \tag{14.18}$$

so the determinant $|p'q'r'|$ will be

$$\begin{aligned}
|p'q'r'| = |pqr| &+ \epsilon_r(x_q - x_p) + \delta_r(y_q - y_p) \\
&+ \epsilon_q(x_p - x_r) + \delta_q(y_p - y_r) \\
&+ \epsilon_p(x_r - x_q) + \delta_p(y_r - y_q) \\
&- \epsilon_r \delta_q + \epsilon_q \delta_r + \epsilon_r \delta_p - \epsilon_p \delta_r - \epsilon_q \delta_p + \epsilon_p \delta_q ,
\end{aligned} \tag{14.19}$$

arranged according to increasing powers of ϵ. All of these determinants $|p'q'r'|$ will have the sign of the first nonvanishing coefficient of (14.19), for all ϵ in the range $0 < \epsilon < \delta$, when δ is sufficiently small. We can therefore set $\epsilon = \frac{1}{2}\delta$, and define pqr true iff $|p'q'r'| > 0$.

This definition makes pqr true iff $|pqr| > 0$ or $|pqr| = 0$ and $p \neq q$, given that $p \prec q \prec r$. For if $|pqr| = 0$ and $p \neq q$, we have $(x_p, y_p) < (x_q, y_q)$ in lexicographic order; hence $x_q - x_p > 0$ or $x_q - x_p = 0$ and $y_q - y_p > 0$, making (14.19) positive. And if $|pqr| = 0$ and $p = q$, there are two subcases. Either $q \neq r$, which makes (14.19) negative since $(x_p - x_r, y_p - y_r) = (x_q - x_r, y_q - y_r) < (0,0)$; or $q = r$, which makes (14.19) assume the negative sign attached to $\epsilon^{(i+j)n+2n-j}$.

We have proved that our ϵ-based definition is equivalent to (14.6). Rule (14.6) therefore defines a CC system; indeed, it defines a realizable CC system—a system that is realizable by points arbitrarily close to the given ones, having no collinear triples. $\square$

Slight changes to (14.16) will produce other ways to define a realizable CC system by slightly perturbing any given set of points. But we have seen that no rule simpler than (14.6) will define a CC system, realizable or not, if we insist that pqr should be true whenever $|pqr|$ is positive, unless repeated points are ruled out.

The lengthy case analysis in the first proof of the theorem indicates that Axiom 5 can come very close to failure in several ways, whenever triples of points are nearly collinear. Therefore it appears very unlikely that any calculation of determinants with less than 100% accuracy will define a ternary relation satisfying Axioms 1–5. Floating-point arithmetic is therefore out of the question unless we restrict coordinates to some domain where floating-point computations are exact. For example, many computer systems now incorporate IEEE standard floating-point arithmetic allowing 53 bits of precision in the C type **double** or the FORTRAN type DOUBLEPRECISION. This means that exact results are obtainable if the input data is first rounded to a fixed-point range of 26 bits or less.

In general, suppose we decide to convert the input data to a fixed-point range of b bits. This means that each x-coordinate is rounded to the nearest value of the form $x/2^{d_x}$, where x is an integer in the range $x_0 \le x < x_0 + 2^b$ and the parameters x_0 and d_x are chosen so that all input data lies between $x_0/2^{d_x}$ and $(x_0 + 2^b)/2^{d_x}$. Each y-coordinate is, similarly, rounded to $y/2^{d_y}$; the values of y_0 and d_y can be independent of x_0 and d_x. Then each determinant $|pqr|$ depends only on the b-bit integers $(x - x_0, y - y_0)$ corresponding to the rounded coordinates $(x/2^{d_x}, y/2^{d_y})$; the values of x_0, d_x, y_0, and d_y have no effect. For programming purposes we can therefore assume that all coordinates (x_p, y_p) are nonnegative integers less than 2^b.

The easiest way to evaluate the determinant $|pqr|$ is to use the formula

$$|pqr| = \begin{vmatrix} x_p & y_p & 1 \\ x_q & y_q & 1 \\ x_r & y_r & 1 \end{vmatrix} = \begin{vmatrix} x_p - x_r & y_p - y_r & 1 \\ x_q - x_r & y_q - y_r & 1 \\ 0 & 0 & 1 \end{vmatrix}$$

$$= (x_p - x_r)(y_q - y_r) - (x_q - x_r)(y_p - y_r), \quad (14.20)$$

which requires exact arithmetic on $2b+1$ bits plus a sign bit. Thus, we can take $b = 26$ if we want to work with standard IEEE double-precision floating-point hardware, or $b = 15$ if we want to use ordinary single-precision integer arithmetic.

This all-integer scheme is attractive even in the case $b = 15$, because we can use rule (14.6) to obtain a CC system that is consistent with respect to the input data except for small perturbations of at most $1/32768 \approx 0.003$ percent of the range. The data in practical problems is probably no more accurate than this. Notice that data values near each other may round to the same integer coordinates, but rule (14.6) is specifically formulated to be appropriate in such situations: Repeated points may occur in the rounded data, but they cause no problem to algorithms based on the CC predicate.

In most applications the determinant $|pqr|$ vanishes only rarely, so it is a waste of time to preprocess the data by sorting it into lexicographic order. Moreover, we have seen that it is advantageous for algorithms to access the data in random order. Thus, we will usually want to implement the test for counterclockwise pqr by using the following procedure, after rounding the data and attaching a unique "serial number" l_p to the point named p:

Step 1. Evaluate the determinant $|pqr|$ with perfect accuracy. If the result is nonzero, return 'true' if it is positive, 'false' if it is negative. Otherwise set $b = $ 'true' and proceed to Step 2.

Step 2. If $l_p > l_q$, interchange $p \leftrightarrow q$ and complement the value of b; if $l_q > l_r$, interchange $q \leftrightarrow r$ and complement the value of b; repeat until $l_p < l_q < l_r$.

Step 3. If $x_p > x_q$, or $x_p = x_q$ and $y_p > y_q$, or $x_p = x_q$ and $y_p = y_q$ and $x_r > x_p$, or $x_p = x_q$ and $y_p = y_q$ and $x_r - x_p$ and $y_r \geq y_p$, complement the value of b.

Step 4. Return the value of b.

If the data is being randomized in order to obtain good expected behavior of an algorithm, the randomization should be independent of the l values. We can usually let l_p be the location where x_p and y_p are stored, if the other data structures point to these coordinate values.

15. Parsimonious algorithms

Let's pause a moment to take stock of where we are. We have studied one of the important primitive operations of computational geometry, the counterclockwise predicate; and we've seen that efficient algorithms for convex hulls in the plane can be designed to use that predicate alone, provided that the implementation of the primitive operation satisfies the axioms of a CC system. The vast majority of CC systems are unrealizable by points in the plane, and our algorithms apply in general: Given any CC system, they will find the unique cycle of extreme points that encloses all other points.

Yet when we studied the problem of actually implementing the counterclockwise predicate for points in the plane, we found that Axiom 5 is difficult to satisfy unless arithmetic is done with perfect accuracy or with careful control over allowable perturbations of the given coordinates. And we know that a ternary predicate can lead to situations incompatible with the basic properties of convex hulls if it does not obey Axiom 5 and the other axioms. Thus our study of CC systems seems to have led to the conclusion that any reliable algorithm for convex hulls, based entirely on the counterclockwise predicate, requires a rather complicated and time-consuming program to implement that predicate.

There are, however, approaches to algorithm design that allow us to work with inexact implementations of primitive operations. For example, Steven Fortune [22] has suggested the term *parsimonious* to describe algorithms that "ask no stupid questions." We can say more precisely that an algorithm is parsimonious with respect to a given system of axioms if the algorithm never evaluates a primitive predicate when the result of that predicate could have been deduced from facts already known because of previously evaluated predicates.

If a parsimonious algorithm works correctly whenever its primitive operations obey the relevant axioms, then it works also when the primitive operations *violate* the axioms—in the sense that the algorithm will always terminate and produce a result consistent with the answers to all questions that its primitive predicates were asked.

We can get some insight into the nature of parsimonious algorithms by studying the problem of sorting from this standpoint. Most algorithms for sorting are based on a primitive operation that compares two keys x and y, returning the value of the predicate $x < y$. For simplicity, we will consider only the case of distinct keys, when the primitive comparison operation is supposed to satisfy two axioms:

R1 (Antisymmetry). $x < y \iff \neg(y < x)$.

R2 (Transitivity). $x < y \wedge y < z \implies x < z$.

(These axioms characterize a transitive tournament.) A parsimonious sorting algorithm will never compare x to y unless both outcomes $x < y$ and $\neg(x < y)$ would be consistent with the results of all previous tests.

Many of the classic sorting algorithms are, in fact, parsimonious with respect to R1 and R2. For example, we will prove below that the following treesort algorithm has the parsimonious property: Start with an empty binary tree, then repeatedly insert $x_1, \ldots, x_N$ into new nodes of that tree, then output the contents of the nodes in symmetric order. To insert x into an empty binary tree, create a new node having key x and make it the root, with empty left and right subtrees. To insert x into a nonempty binary tree with key y in its root node, compare x to y; then recursively insert x into the left or the right subtree according as $x < y$ or not. This algorithm clearly terminates and outputs the original keys in some order $x_{p(1)} x_{p(2)} \cdots x_{p(N)}$ such that $p(1)p(2)\ldots p(N)$ is a permutation of $\{1, 2, \ldots, N\}$ and such that

$$x_{p(1)} < x_{p(2)} < \cdots < x_{p(N)}, \tag{15.1}$$

whenever Axioms R1 and R2 hold.

Merge sorting is another algorithm that turns out to be parsimonious. This one uses divide and conquer: Given a list of N keys $(x_1, \ldots, x_N)$, if $N \leq 1$ there is nothing to do; otherwise sort the two sublists $(x_1, \ldots, x_{\lfloor N/2 \rfloor})$ and $(x_{\lfloor N/2 \rfloor + 1}, \ldots, x_N)$ recursively, then merge these to form a final list. To merge $(x_1, \ldots, x_m)$ with $(y_1, \ldots, y_n)$, producing a sorted list $(z_1, \ldots, z_{m+n})$, let $(z_1, \ldots, z_m) = (x_1, \ldots, x_m)$, if $n = 0$; or let $(z_1, \ldots, z_n) = (y_1, \ldots, y_n)$, if $m = 0$; or set $z_1 = x_1$ and $(z_2, \ldots, z_{m+n}) = (x_2, \ldots, x_m)$ merged with $(y_1, \ldots, y_n)$, if $m > 0$, $n > 0$, and $x_1 < y_1$; or set $z_1 = y_1$ and $(z_2, \ldots, z_{m+n}) = (x_1, \ldots, x_m)$ merged with $(y_2, \ldots, y_n)$, otherwise. Again, if the $<$ operation obeys R1 and R2, merge sorting produces a permutation satisfying (15.1).

In order to prove that treesort is parsimonious, we want to show that whenever the algorithm tests whether or not $x < y$, there is a model of the input consistent with the test going either way. One way to construct such models is to assign real values x'_k in the open interval $(0 .. 1)$ to each input x_k, in such a way that $x'_k < y'$ whenever the algorithm finds $x_k < y$ true, $x'_k > y'$ whenever the algorithm finds $x_k < y$ false. As the algorithm proceeds, this means x'_k will be constrained to lie in an interval $(l_k .. r_k)$, where l_k is initially 0 and r_k is initially 1. When the algorithm inserts x_k into an empty binary tree, we can for example set $x'_k = \frac{1}{2}(l_k + r_k)$. Then when the algorithm compares x_k to y, we will know by induction that y' is the midpoint of the interval $(l_k .. r_k)$, so we can set $r_k \leftarrow y'$ if $x < y$, otherwise $l_k \leftarrow y'$, and proceed recursively into the appropriate subtree. This rule gives x'_k a terminating binary expansion with bits 0 or 1 to indicate a left or right branch in the tree, ending with a 1 bit when x_k is inserted. None of the tests $x < y$ could be redundant with respect to R1 and R2, because the relation $<$ on real numbers satisfies those axioms and the numbers $x'_1, \ldots, x'_N$ cause the algorithm to take that branch.

Another way to show that treesort is parsimonious goes backward instead of forward: Imagine playing the algorithm out with all possible combinations of true and false whenever a comparison is made; this means we are essentially viewing the algorithm as a huge decision tree. We want to show that each of the leaves of that tree—corresponding to a terminating computation—has a model $(x'_1, \ldots, x'_N)$ satisfying R1 and R2, which causes the algorithm to reach that leaf. We can compute the numbers $(x'_1, \ldots, x'_N)$ by taking $x'_{p(1)} \ldots x'_{p(N)} = 1 \ldots N$, where $p(1) \ldots p(N)$ is determined by symmetric order of the binary tree constructed on the path to a given leaf. In this way, for example, x'_1 will be equal to k iff the binary tree ends up with $k - 1$ nodes in its left subtree. We don't know the model values $x'_1, \ldots, x'_N$ until the algorithm runs to completion, but then we can assign them with hindsight.

A backward analysis seems to be the easiest way to prove that merge sort is parsimonious. We simply need to find a model corresponding to every leaf of the decision tree so that the result of sorting will be $x'_{p(k)} = k$; in other words, we want to run the merge sort backwards to find original numbers $x'_k = p^{-1}(k)$ that cause it to follow a given path. If the numbers $\{z_1, \ldots, z_{m+n}\}$ after merging come from a given set Z, and if the merging took a particular computation path, we can reconstruct the sets $X = (x_1, \ldots, x_m)$ and $Y = (y_1, \ldots, y_n)$ that led to this path by playing the recursion in reverse. Elegant formulations of such proofs can no doubt be worked out if the idea of parsimonious algorithms proves to be important; further details are left to the reader.

Now suppose we have a relation $<$ that does *not* obey R1 and R2. For example, the relation might be based on a majority vote among $2m + 1$ observers, who are asked to rank objects according to some criterion; or it might ask whether $x(t)$ is less than $y(t)$ at some time t, when x and y are the solutions to differential equations that can only be solved numerically with limited precision; or it might be any relation whatever. When a parsimonious sorting algorithm is applied to keys $x_1, \ldots, x_N$ governed by an arbitrary relation $<$, the algorithm has no way of knowing that Axioms R1 and R2 are not satisfied; so it finds a permutation $p(1) \ldots p(N)$ of $\{1, \ldots, N\}$ such that (15.1) is deducible from the queries made.

A parsimonious sorting algorithm will always find a permutation that satisfies

$$\text{either} \quad x_{p(j)} < x_{p(j+1)} \quad \text{or} \quad \neg(x_{p(j+1)} < x_{p(j)}), \quad \text{for } 1 \le j < N. \tag{15.2}$$

For the permutation (15.1) is unique, in the presence of R1 and R2. And the algorithm could not be correct unless it had explicitly compared $x_{p(j)}$ with $x_{p(j+1)}$; otherwise it wouldn't know which was the smaller. Hence the result of comparing $x_{p(j)}$ to $x_{p(j+1)}$ must be as stated in (15.2); anything else would contradict (15.1) by R1.

These observations have several consequences. First, any sorting algorithm that is parsimonious with respect to R1 and R2 will produce an output satisfying (15.1), if the relation $<$ satisfies R1 but is not necessarily transitive. (This is a constructive version of the well known theorem that any tournament on N vertices contains a directed path of length N.) Furthermore, any such sorting algorithm produces outputs satisfying

$$x_{p(1)} \le x_{p(2)} \le \cdots \le x_{p(N)}, \tag{15.3}$$

where $x \le y$ means $\neg(y < x)$, if its relation satisfies not R1 but the weaker law

R1′ (Asymmetry). $x < y \implies \neg(y < x)$.

Hence any such algorithm, parsimonious for unequal keys, will also sort when equal keys are present.

The concept of parsimony depends strongly on the axioms being considered. For example, suppose we replace R1 and R2 by R1′ and

R2′ (Cotransitivity). $\neg(x < y) \land \neg(y < z) \implies \neg(x < z)$.

Then R1′ and R2′ imply R2, but they are not strong enough to imply R1 because the relation $x < y$ might never be true. The following algorithm for sorting three elements $\{x, y, z\}$ is parsimonious with respect to R1′ and R2′:

```
if (x < y)
        if (y < z) output (xyz)                          x < y < z
        else if (z < y)
                if (x < z) output (xzy)                  x < z < y
                else output (zxy)                        z ≤ x < y
        else output (xzy)                                x < y ≤ z ≤ y
else if (y < x)
        if (x < z) output (yxz)                          y < x < z
        else if (y < z) output (yzx)                     y < z ≤ x
                else  output (zyx)                       z ≤ y ≤ x
        else if (x < z) output (xyz)                     x ≤ y ≤ x < z
                else output (zyx).                       z ≤ x ≤ y ≤ x.
```

In all cases, it produces a permutation $x'y'z'$ of $\{x, y, z\}$ that would satisfy $x' \leq y' \leq z'$ if R1' and R2' hold. (We need to observe, for example, that

$$x < y \land y \leq z \Longrightarrow x < z,$$

since $y \leq z$ and $z \leq x$ implies $y \leq x$, i.e., $\neg(x < y)$.) But its output does not always satisfy (15.2) in the presence of an arbitrary relation. For example, if we have $x < z$ and $z < y$ and no other true cases, the algorithm outputs xyz but we have neither $y < z$ nor $\neg(z < y)$. Therefore being parsimonious with respect to R1' and R2' is not as strong as being parsimonious with respect to R1 and R2.

If a parsimonious algorithm is known to have guaranteed efficiency in the worst case, when presented with data that satisfies the relevant axioms, it will be just as efficient in the worst case when presented with *arbitrary* data. Thus, for example, merge sort will always find a permutation satisfying (15.2) in $O(N \log N)$ time. But such performance guarantees do not necessarily hold if we know only that a parsimonious algorithm is efficient on the average when it *randomizes* its data. For example, treesort is known to make exactly

$$T_N = 2(N + 1) H_N - 4N \tag{15.4}$$

comparisons on the average if it inserts N distinct keys in random order; but if the relation $x < y$ is false for all x and y, treesort will make $\frac{1}{2} N(N - 1)$ comparisons regardless of the order in which it chooses to insert the keys.

Incidentally, we can show that treesort does make at most T_N comparisons on the average if the relation $<$ satisfies R1 (i.e., if the keys form a tournament), and if the keys are inserted in random order. The proof is by induction on N. Assuming the truth of this assertion for tournaments of size less than N, we can conclude that the average number of comparisons on an N-tournament will be at most

$$N - 1 + \sum_x \frac{1}{N} \left(T_{s(x)} + T_{N-1-s(x)} \right), \tag{15.5}$$

where $s(x)$ is the score of x. (Here x denotes the key of the root node; each root key occurs with probability $1/N$.) If two vertices x and y have the same score s, with say $x < y$, we can change the relation to $y < x$, thereby changing their respective scores to $s - 1$ and $s + 1$. Then (15.5) increases by $1/N$ times

$$T_{s-1} - 2T_s + T_{s+1} + T_{N-2-s} - 2T_{N-1-s} + T_{N-s}$$
$$= (T_{s+1} - T_s) - (T_s - T_{s-1}) + (T_{N-s} - T_{N-1-s}) - (T_{N-1-s} - T_{N-2-s}),$$

which is nonnegative because $T_{n+1} - T_n = 2(H_{n+1} - 1)$ is an increasing function of n. It follows that (15.5) is at most

$$N - 1 + \sum_{s=0}^{N-1} \frac{1}{N} \left(T_s + T_{N-1-s} \right),$$

which is just T_N. We have essentially proved that treesort makes even fewer than T_N comparisons, on the average, when it is presented with a nontransitive tournament.

Now let's consider convex hulls instead of sorting. The simple hull insertion algorithm of section 13 is easily seen to be parsimonious with respect to Axioms 1–5; therefore it will always terminate with something hull-like, even if the counter-clockwise predicate is implemented sloppily. But hull insertion has a bad worst-case running time, so its parsimonious nature doesn't help us much. One of the other algorithms proves to be more interesting:

Theorem. *The treehull algorithm of section 11 is parsimonious with respect to the axioms of a realizable CC system.*

Proof. We must show that any computation path taken by the algorithm has a model in the plane; some set of points should make it assume each of its possible behaviors. We can construct such a set by running the algorithm backwards, as we did for merge sorting.

Suppose the algorithm terminates with the m-point convex hull $(t_1, \ldots, t_m)$, and with a given binary search tree on $t_2, \ldots, t_m$. Then our model will contain the points $t'_1, \ldots, t'_m$, chosen to lie on any convex m-gon; for concreteness, we can let them be equally spaced points on a circle. Now we move a step backward. If the last point p considered caused the algorithm to enter Case 1, then p was inserted into the convex hull and it may also have replaced previous points $p_1, \ldots, p_r$, where $r \geq 0$. To undo this step, we remove p' from our m-gon, and insert $p'_1, \ldots, p'_r$ between its two former neighbors in such a way that the resulting polygon is still convex. This is clearly always possible. We also restore the previous state of the binary search tree. The algorithm would then take the prescribed path if we reinserted p' and ran it forward. (Note that the algorithm does not make the "final" counterclockwise test that would have eliminated too much of the convex hull if it had succeeded; if Case 1 finds pt_1t_m and pt_2t_1 and $\cdots$ and $pt_{m-2}t_{m-1}$, it "knows" that $pt_{m-1}t_m$ must be false, so it does not make the test.)

On the other hand if the last point p led the algorithm to Case 2, point p may or may not have entered the hull. If not, it was found to lie in $\triangle t_1 t_{k-1} t_k$, and we can choose p' to be any point in that triangle. And if p did join the hull in Case 2, then p' is one of the points $t'_2, \ldots, t'_m$. In this case we move backward as in Case 1, by removing p' and inserting $r \geq 0$ points $p'_1, \ldots, p'_r$ inside the polygon between the former neighbors of p'; we must also do this in such a way that some of the points p'_j lie to the left of $t'_1 p'$ and some lie to the right, depending on the value of k found by the algorithm in its tree search. Again, this is always possible, and it causes the algorithm to take exactly the computation path desired.

Therefore we can run the algorithm in reverse until all of its input points have been modeled by points in the plane. Those points (x_k, y_k) will cause the algorithm to go through the same motions, when running forward. $\square$

This theorem has a certain appeal for the "programmer on the street." Treehull is a rather simple algorithm, and it can be implemented with splay trees or treaps to obtain reasonably good worst-case or average-case bounds on its running time. (We have observed that randomization over input data does not automatically preserve performance estimates of parsimonious algorithms; but randomization within the data structure does.)

The theorem suggests that a straightforward floating-point implementation of the counterclockwise predicate will be satisfactory for practical purposes, because deficiencies in satisfying the axioms cannot cause the treehull algorithm to "blow up." Moreover, our intuition tells us that a slightly inaccurate implementation should not be terribly misleading to the rest of the algorithm; any mistakes should be essentially "reasonable," in the sense that small perturbations to the input data would probably be consistent with the algorithm's behavior.

On the other hand, the knowledge that an algorithm is parsimonious is only a weak indication of robustness. The treehull algorithm might, for example, produce a "convex hull" $(t_1, \ldots, t_m)$ that does not actually form a convex polygon, if it is based on a potentially unreliable counterclockwise test. One pass around $(t_1, \ldots, t_m)$ to remove concave junctions will cure that problem, but still we would prefer to have some sort of quantitative guarantee of stability within a certain tolerance. And even if such a guarantee is found, the resulting hull will not be uniquely determined; modifications to the order in which points are processed, or changes in the internal randomization of data structures, will produce different "convex hulls" on different runs, when the counterclockwise predicate is not a bona fide CC system, just as treesort will find different paths of length N in a nontransitive tournament when it processes the vertices in different orders.

From these considerations it appears that the degeneracy-breaking methods of section 14 are preferable to the use of parsimonious algorithms, even when quantitative estimates for the error inflation of the parsimonious algorithms are available. If we start by rounding all the data to some fixed-point range, thereby perturbing the input points by at most a known, small percentage of their total range, we can proceed from there to construct a totally rigorous CC system. The resulting algorithm is substantially simpler than all other approximation-based algorithms that have been proposed so far [22, 37, 42, 55], and it has the additional advantage of uniqueness. The same convex hull will necessarily be found by any algorithm that finds the convex hull of a CC system, given a definite way to do the fixed-point rounding and to break ties as explained at the end of section 14.

Incidentally, the daghull algorithm of section 12 is not parsimonious. For example, consider the instructions (12.3) compiled in Step H4. If the very next point p' comes through these instructions and finds first $p'sp$, then $p'tp$, the algorithm in Step H3 will immediately make a redundant test of $p'ps$. Alternatively, if we have $p'ps$ and $p'tp$, any subsequent point p'' might find $p''sp$ and $p''tp$, after which the algorithm will make a test of $p''p's$ which it could have deduced is false. There is no obvious modification that will make the algorithm parsimonious without also making it substantially more complicated.

The results of section 6 above imply that the general task of taking an arbitrary algorithm and making it parsimonious with respect to CC systems is NP-complete; however, there may be a way to avoid this complexity in specific algorithms such as daghull.

16. Composition of CC systems*

It is possible to combine small CC systems into larger ones using constructions that are analogous to wreath products and Cartesian products of algebraic systems and graphs. The purpose of this section is to outline one such construction, in case it should turn out some day that CC systems have applications unrelated to orientations of points in the plane.

These constructions are motivated by the reflection networks considered in section 8 above. Let us say that a linear ordering $p \succ q$ of the points of a CC system $\mathcal{C}$ is a *projective order* if it is one of the orderings that can occur in a sequence of $\binom{n}{2}$ ordered pairs that defines $\mathcal{C}$ and obeys the betweenness rule (7.1). These are the orderings with the property that we can extend $\mathcal{C}$ to a CC system on one further point ∞, with the rule that ∞pq holds iff $p \succ q$. In a realizable CC system, they include the orderings that can occur if all points are projected onto a straight line that is not perpendicular to the direction between any two points. Projective orders are also the orderings of points from top to bottom, at the left of the reflection networks in section 8. They correspond to the cutpaths considered in section 9, where we proved that at most 3^n projective orders are possible in a CC system. It is not difficult to verify that a linear ordering $\succ$ is a projective order if and only if the following laws hold:

$$\neg\, (p \succ t \,\wedge\, q \succ t \,\wedge\, r \succ t \,\wedge\, t \in \Delta pqr)\,; \tag{16.1}$$

$$\neg\, (q \succ t \,\wedge\, t \succ s \,\wedge\, t \succ r \,\wedge\, tsq \,\wedge\, tqr \,\wedge\, tsr)\,. \tag{16.2}$$

(Rule (16.1) is Axiom 5 with $s = \infty$; rule (16.2) is Axiom 5 with $p = \infty$. Note that these rules are special cases of (14.11) and (14.14), which we deduced for lexicographic order in the plane; indeed, lexicographic order is a projective order.)

Now suppose $\mathcal{C}$ is a "master" CC system on a set of points C, and suppose that there is a subsidiary CC system $\mathcal{C}_p$ on disjoint sets of points C_p for each $p \in C$. Let $\succ$ be a projective order on $\mathcal{C}$ and let $\succ_p$ be a projective order on $\mathcal{C}_p$. Then we obtain a CC system on the set of all points

$$(p, p'), \quad p' \in C_p \tag{16.3}$$

by saying that $(p, p')(q, q')(r, r')$ holds iff $\{p, q, r\}$ are distinct and pqr is true in $\mathcal{C}$; or $p = q = r$ and $p'q'r'$ holds in $\mathcal{C}_p$; or

$$
\begin{aligned}
(p = q \succ r \,\wedge\, q' \succ_p p') &\vee (r \succ p = q \,\wedge\, p' \succ_p q') \\
\vee\, (q = r \succ p \,\wedge\, r' \succ_q q') &\vee (p \succ q = r \,\wedge\, q' \succ_q r') \\
\vee\, (r = p \succ q \,\wedge\, p' \succ_r r') &\vee (q \succ r = p \,\wedge\, r' \succ_r p')\,.
\end{aligned}
\tag{16.4}
$$

The corresponding reflection network is obtained from a network for $\mathcal{C}$ and by first replacing each line containing a point p by a set of lines containing (p, p') with

---

* This section is independent of the remainder of the monograph and can be omitted without hurting the author's feelings.

$p' \in C_p$, and by replacing each transposition module by a sequence of transpositions that interchanges two sets of lines. For example, the sequence

$$(16.5)$$

interchanges the two lines (a, b) with the three lines (c, d, e). Then we append the reflection networks for C_p and $\succ_p$ at the right, to reflect the points (p, p'). The resulting network clearly reflects all points (16.3) by making adjacent transpositions, so the corresponding ternary predicate must satisfy Axioms 1–5.

More general constructions are possible when the reflections of C_p are interspersed with the interchange operations, instead of being done at the end; but then the rules become even more complex than (16.4). A pleasant special case occurs when all the systems C_p are isomorphic. When C and C_p are also realizable, it corresponds to replacing each point in the plane by a tiny cluster of other points, something like satellites within galaxies.

17. The incircle predicate

Suppose we are given a set of points in the plane such that no three are collinear and no four lie on a circle. Then the *Delaunay triangulation* [13] is well-defined; this is the set of positively oriented triangles pqr whose circumcircles contain no other points. The Delaunay triangulation makes it easy to compute the *Voronoi diagram* [69, 59] of the points, the regions in which a given point is closest, because the edges of the Voronoi diagram are perpendicular bisectors of the edges of the Delaunay triangulation. The latter edges may be said to connect "neighboring" points, in the sense that two points are neighbors iff their Voronoi regions are adjacent.

It is well known that the point s lies inside the circle passing through the vertices $\{p, q, r\}$ if and only if the determinant

$$\det \begin{pmatrix} x_p & y_p & x_p^2 + y_p^2 & 1 \\ x_q & y_q & x_q^2 + y_q^2 & 1 \\ x_r & y_r & x_r^2 + y_r^2 & 1 \\ x_s & y_s & x_s^2 + y_s^2 & 1 \end{pmatrix} \tag{17.1}$$

has the same sign as the determinant $|pqr|$ of (1.1). We shall write $|pqrs|$ for the determinant (17.1), and we will consider the quaternary predicate $pqrs$ to be true iff $|pqrs| > 0$; this is the "InCircle" predicate of Guibas and Stolfi [38]. The Delaunay triangulation consists of those triangles with pqr true and with $pqrs$ false for all other points s.

It is convenient to extend the plane by introducing a point ∞ such that $\infty qrs = qrs$. This definition makes sense if we consider the behavior of $|pqrs|$ as x_p and/or y_p approach infinity in (17.1), because the sign of the determinant will approach the sign of the cofactor of $x_p^2 + y_p^2$, which is $|qrs|$. In these terms, Δpqr belongs to the Delaunay triangulation iff $spqr$ holds for all $s \notin \{p, q, r\}$, including $s = \infty$.

Lemma. *If s is any point of the plane, there is a mapping $p \mapsto p^s$ of all the other points such that $spqr$ holds iff $|p^s q^s r^s| > 0$.*

Proof. If p has coordinates (x_p, y_p), let

$$p^s = \left((x_p - x_s)/\Delta_{ps}^2,\ (y_s - y_p)/\Delta_{ps}^2\right), \qquad \Delta_{ps}^2 = (x_p - x_s)^2 + (y_p - y_s)^2. \quad (17.2)$$

Then we have

$$|spqr| = \det \begin{pmatrix} x_s & y_s & x_s^2 + y_s^2 & 1 \\ x_p & y_p & x_p^2 + y_p^2 & 1 \\ x_q & y_q & x_q^2 + y_q^2 & 1 \\ x_r & y_r & x_r^2 + y_r^2 & 1 \end{pmatrix}$$

$$= \det \begin{pmatrix} 0 & 0 & 0 & 1 \\ x_p - x_s & y_p - y_s & (x_p - x_s)^2 + (y_p - y_s)^2 & 1 \\ x_q - x_s & y_q - y_s & (x_q - x_s)^2 + (y_q - y_s)^2 & 1 \\ x_r - x_s & y_r - y_s & (x_r - x_s)^2 + (y_r - y_s)^2 & 1 \end{pmatrix}$$

$$= -\det \begin{pmatrix} x_p - x_s & y_p - y_s & \Delta_{ps}^2 \\ x_q - x_s & y_q - y_s & \Delta_{qs}^2 \\ x_r - x_s & y_r - y_s & \Delta_{rs}^2 \end{pmatrix} = |p^s q^s r^s| \Delta_{ps}^2 \Delta_{qs}^2 \Delta_{rs}^2 \quad (17.3)$$

by column and row operations on determinants, and the lemma follows since we are assuming that $\Delta_{ps}^2 \Delta_{qs}^2 \Delta_{rs}^2 > 0$.

If we let $\infty^s = 0$, the stated result holds also when p, q, or r is infinite. For $s\infty qr$ is true iff ∞sqr is false iff $|sqr| < 0$, and we have

$$|\infty^s q^s r^s| = \det \begin{pmatrix} x_q - x_s & y_s - y_q \\ x_r - x_s & y_s - y_r \end{pmatrix} \Delta_{qs}^2 \Delta_{rs}^2 = -|sqr| \Delta_{qs}^2 \Delta_{rs}^2$$

by (14.18). □

Notice that if we represent the point $p = (x_p, y_p)$ by the complex number $z_p = x_p + i y_p$, the image p^s defined in (17.2) is simply

$$\overline{(z_p - z_s)}/|z_p - z_s|^2 = 1/(z_p - z_s).$$

Thus $spqr$ is true if and only if the points $1/(z_p - z_s)$, $1/(z_q - z_s)$, $1/(z_r - z_s)$ have a counterclockwise orientation.

The lemma reduces the incircle test for points in the plane to a counterclockwise test. This property is, in fact, strong enough to imply that an abstract set of points possesses an analog of the Delaunay triangulation. Let us define a *CCC system* to be any quaternary predicate $spqr$ on distinct points, having the property that for each point s the ternary predicate $spqr$ defined for all distinct points $p, q, r \neq s$ is a CC system. We also require that $spqr$ is true iff $pqrs$ is false. The *Delaunay triangulation* of any CCC system can then be defined as the set of all "triangles" Δpqr such that $spqr$ holds for all $s \notin \{p, q, r\}$.

The axioms for a CCC system are therefore a simple extension of the axioms for a CC system. We can restate them as follows:

Axiom C1. $pqrs \implies \neg spqr$.

Axiom C2. $pqrs \implies \neg pqsr$.

Axiom C3. $pqrs \lor pqsr$.

Axiom C4. $pqrt \land prst \land psqt \implies pqrs$.

Axiom C5. $utsp \land utsq \land utsr \land utpq \land utqr \implies utpr$.

Here Axioms C2, C3, C4, C5 are like Axioms 2, 3, 4, 5 but with another point adjoined; Axiom C1 is a modification of Axiom 1.

In section 20 we will consider an axiom that is weaker than C4, leading to a relation on quadruples that corresponds to convex hulls in three dimensions. The stronger axiom C4 stated here will turn out to represent the special case of points in 3D that lie on the surface of a sphere.

We can prove as before that Axioms C1, C3, C4, and C5 are independent; but it turns out that C2 is a consequence of C1 and C3. Indeed, $pqrs \implies \neg spqr \implies sprq \implies \neg qspr \implies qsrp \implies \neg pqsr$ if we apply C1 and C3 in alternation. From C1 and C3 we can in fact deduce that every transposition of two elements negates the value of $pqrs$. For example, to show that $pqrs$ implies $\neg psrq$, we have

$$pqrs \implies \neg spqr \implies sprq$$
$$\implies \neg prqs \implies prsq \implies \neg qprs \implies qpsr$$
$$\implies \neg psrq\,;$$

the third and seventh of these implications use the contrapositive of C1, which can be written

$$pqrs \implies \neg qrsp\,. \tag{17.4}$$

In applications it is sometimes helpful to note that we can replace (p, q, r, s, t) by (t, r, q, s, p) in C4 and rearrange the order of variables within quadruples to get

$$pqrt \land prst \land psqt \implies qrst\,. \tag{17.5}$$

Thus, we can sometimes deduce the values of all five quadruples on $\{p, q, r, s, t\}$ when we know only three of them. In the presence of Axioms C1 and C3, Axiom C4 is equivalent to the following cyclically symmetrical statement: "Exactly two or three of the quadruples $pqrs$, $qrst$, $rstp$, $stpq$, $tpqr$ are true."

Theorem. *Every point of a CCC system on four or more points is a vertex of at least three triangles of the Delaunay triangulation. The triangles with vertex p form a cycle,*

$$\Delta pt_1t_2\,, \ \Delta pt_2t_3\,, \ \ldots\,, \ \Delta pt_{m-1}t_m\,, \ \Delta pt_mt_1\,. \tag{17.6}$$

Proof. By definition, Δpqr is part of the Delaunay triangulation iff $\neg pqrs$ for all $s \notin \{p, q, r\}$. Let $\alpha_p(q, r, s) = \neg pqrs$; this ternary relation α_p defines a CC system on the points other than p, because it is obtained by complementing the triples of the CC system obtained by fixing p. Therefore Δpqr is part of the Delaunay triangulation iff qr is in the convex hull of the CC system α_p, and the result follows immediately from the theorem of section 11. $\square$

One consequence of this theorem is that every triangle Δpqr in the Delaunay triangulation of a CCC system has three neighboring triangles of the forms $\Delta qpr'$, $\Delta rqp'$, and $\Delta prq'$. In other words, every Delaunay edge is part of exactly two Delaunay triangles. Therefore we can represent the triangulation with a rather simple data structure, related to the quad-edge technique of [38]. Indeed, a slight extension of the proof above shows that even more is true: The two-dimensional manifold defined by the Delaunay triangulation of any CCC system is orientable, in fact homeomorphic to a sphere. This will be a consequence of the incremental algorithm in the following section.

From an intuitive standpoint it is perhaps easiest to visualize triangulations on the surface of a sphere, rather than in the plane. Consider the mapping that takes point (x, y) into (ξ, η, ζ), where

$$\xi = \frac{2x}{x^2 + y^2 + 1}, \quad \eta = \frac{2y}{x^2 + y^2 + 1}, \quad \zeta = \frac{x^2 + y^2 - 1}{x^2 + y^2 + 1}; \qquad (17.7)$$

then $\xi^2 + \eta^2 + \zeta^2 = 1$, so (ξ, η, ζ) is a point on the unit sphere. Conversely, any such point has an inverse image (x, y) given by

$$x = \frac{\xi}{1 - \zeta}, \quad y = \frac{\eta}{1 - \zeta}, \qquad (17.8)$$

except for the "north pole" $(\xi, \eta, \zeta) = (0, 0, 1)$ which corresponds to ∞. Elementary manipulation of determinants shows that $|spqr|$ has the same sign

$$\det \begin{pmatrix} s_\xi & s_\eta & s_\zeta & 1 \\ p_\xi & p_\eta & p_\zeta & 1 \\ q_\xi & q_\eta & q_\zeta & 1 \\ r_\xi & r_\eta & r_\zeta & 1 \end{pmatrix}, \qquad (17.9)$$

where $p' = (p_\xi, p_\eta, p_\zeta)$ corresponds to (p_x, p_y); this determinant is the volume of the tetrahedron whose vertices are p', q', r', and s', so it is positive for all s' and for fixed p', q', r' if and only if $p'q'r'$ is a face of the three-dimensional convex hull when the given points have been projected onto the unit sphere.

Any CC system might arise by fixing a point of a CCC system. The following construction, due to Günter Ziegler [72], shows how to obtain a CCC that extends an arbitrarily given CC system $\mathcal{C}$: Number the vertices of $\mathcal{C}$ in such a way that 1 is in the convex hull, then 2 is in the convex hull of the points remaining when 1 is removed, and so on; thus vertex k will be an extreme point of the CC system on

$\{k, k + 1, \ldots, n\}$. Also adjoin an additional point 0. Then if $p < q < r < s$, let $pqrs = qrs$. (It follows that $pqrs = qrs$ whenever $p = \min(p, q, r, s)$.)

Regardless of how we number the vertices, we can show that the tournament obtained from this construction by fixing any two points is vortex-free. Suppose we fix $\{a, b\}$, where $a < b$, and suppose the corresponding tournament defined by $abxy$ contains a vortex on $\{p, q, r, s\}$ where $p < q < r < s$. We cannot have such a vortex when $a < p$, because $abxy = bxy$ for all $x, y \in \{p, q, r, s\}$ in that case. If $p < a < q$, then $abpx = pabx = abx = b\bar{a}x$ and $abxy = bxy$ for all $x, y \in \{q, r, s\}$; so a vortex on $\{p, q, r, s\}$ would imply a vortex on $\bar{a}, q, r, s\}$. Therefore we must have $q < a$. But then $abpx = abqx$ for $x \in \{r, s\}$, making a vortex impossible.

To complete the proof we must show that C4 holds. If it fails, we have $pqrt$, $prst$, $psqt$, and $psrq$, for some vertices p, q, r, s, and t. Axiom 4 holds in $\mathcal{C}$, so we cannot have $p = \min(p, q, r, s, t)$. By symmetry we can therefore assume that $q = \min(p, q, r, s, t)$. This implies ptr, pst, and prs in $\mathcal{C}$; hence the relation $prst$ implies that $p = \min(p, r, s, t)$. But this contradicts the principle by which we numbered the vertices, since $p \in \Delta rst$.

18. A generalized Delaunay algorithm

Let us now consider an efficient way to find the Delaunay triangulation of any given CCC system. Our approach will be incremental as in the algorithms for convex hulls in sections 11–13 above.

Suppose we have found the Delaunay triangulation of all points except p. Then it is easy to characterize the Delaunay edges pt_k of (17.6): A point t is part of some Delaunay triangle $\Delta ptt'$ with respect to all points if and only if it is part of some Delaunay triangle Δtqr with respect to allpoints except p, where $tqrp$ holds. Consider the triangles Δtqr involving t when p is excluded; these have the defining property $stqr$ for all $s \notin \{t, q, r, p\}$. If $ptqr$ is true as well, triangle Δtqr will be part of the overall triangulation; so the edge pt cannot be Delaunay unless $tqrp$ holds for some Δtqr. And in the latter case, p must be an extreme point of the CC system we obtain by fixing point t, since p lies outside the convex hull defined by the other points of that CC system.

We can therefore formulate a naïve algorithm for Delaunay triangulation in general. To add a new point p, just run through all triangles Δtqr of an existing triangulation, and mark all the triangles such that $tqrp$ holds. The new triangulation consists of all unmarked triangles plus all edges pt where t is a vertex of a marked triangle. There will always be a way to arrange these edges pt_k into a cycle like (17.6); the triangles $\Delta pt_k t_{k+1}$ now take the place of the triangles previously marked.

In order to make this algorithm efficient, we need a good way to locate a single marked triangle. Once we've found Δtqr with $tqrp$ true, we know that pt, pq, and pr will be part of the cycle of new Delaunay edges, and it will be a simple matter to find the full cycle by looking at triangles adjacent to triangles already marked.

The incircle predicate $spqr$ generally involves nontrivial computation. So we will assume that one of the points of our CCC system is called ∞, and we will design our algorithm so that most of the incircle tests it makes have the form ∞pqr. Tests

of the latter kind are simply counterclockwise predicates in a CC system, and we
may assume that such tests are less expensive than the evaluation of *spqr* in general.
The introduction of ∞ makes the algorithm less symmetric, hence more complicated,
but it has important practical consequences when we wish to compute the Delaunay
triangulation of points in the plane. Indeed, the algorithm to be described may well
be the fastest and most easily implemented of all methods currently known for that
problem. It is a simplification, modification, and generalization of the algorithm in
section 3 of [36].

We will call the algorithm *dag triangulation*, because it is based on a binary
branching structure that forms a directed acyclic graph, similar to the "compiled
instructions" in the daghull algorithm of section 12 above. (The reader is advised to
review daghull before proceeding further.) Besides the instructions, there is a data
structure of *arcs*, representing the current triangulation. There will be $6n - 6$ arcs
altogether when the algorithm has found the triangulation of n points besides ∞.

Each arc has a unique *mate*. One convenient way to represent this inside a
computer is to arrange things so that the mate of the arc in position k appears in
position $6N - 7 - k$, for $0 \le k < 6N - 6$, using an array of $6N - 6$ arcs when N
is the total number of noninfinite points. We will call these arcs $a_1, a_2, \ldots, a_{3N-3}$,
$b_{3N-3}, \ldots, b_2, b_1$, respectively, so that a_j and b_j are always mates. The noninfinite
points are called *vertices*; if p, q, r, s are any distinct vertices, the algorithm is supposed
to be able to compute the counterclockwise predicate $pqr = \infty pqr$ and the incircle
predicate *spqr*.

An arc a conceptually has a triangle on its left and a mate on its right. It has
three fields:

vert(a) points to the vertex this arc leads to, or is Λ (the null pointer) if the arc
   leads to ∞;

next(a) points to the next arc having the same triangle at its left;

inst(a) points to the "terminal node" for that triangle, as explained below.

We have $next\big(next(next(a))\big) = a$ and $inst\big(next(a)\big) = inst(a)$ for all a. If arc a runs
from vertex u to vertex v, we have vert(a) = v and vert(mate(a)) = u.

The dag is an array of instruction nodes, similar to the nodes in section 12,
but there are two differences: (1) Each node now has four fields (p, q, α, β) instead
of two. (2) The total number of nodes is not known in advance, so nodes should
be allocated dynamically. The first two fields p, q usually point to vertices; the other
fields α, β point to other nodes. The instruction represented by node (p, q, α, β) means,
intuitively, "if v lies to the left of pq then goto α else goto β." By starting at the root
of the dag and following instructions, we will be able to find the triangle containing
a vertex v that we wish to insert. Eventually we will reach a *terminal instruction*,
which is a node having the special form $(\Lambda, a, -, -)$. There is one terminal instruction
node for each triangle in the current triangulation. The first field is Λ, to distinguish
terminal nodes from branch nodes. The second field, a, points to one of the arcs of the
corresponding triangle; *inst(a)* points back to the terminal node $(\Lambda, a, -, -)$. The α
and β fields of terminal nodes are not used, but they are present in case the terminal

node later becomes nonterminal. If the triangle has ∞ as an endpoint, arc a will be the arc with $vert(a) = \Lambda$; otherwise a might be any one of the triangle's three arcs.

If $vert(a) \neq \Lambda$, the triangle Δqrs represented by a terminal node will be the set of points p such that we have $\infty pqr \wedge \infty prs \wedge \infty psq$; if p satisfies this condition, we have $qrsp$ by (17.5). If $vert(a) = \Lambda$, the "triangle" $\Delta \infty rs$ represented by a terminal node will actually be the "wedge" $\angle \bar{r}'rs$ consisting of all points p such that ∞prs and $\infty prr'$ holds, where r', r, s are consecutive elements t_{j+1}, t_j, t_{j-1} of the current convex hull. Every point p seen so far satisfies $\infty pt_{j-1}t_j$ for $1 \leq j \leq m$, if we let $t_0 = t_m$; every point p to be inserted either lies inside one of the triangles Δqrs or in a unique wedge $\angle \bar{t}_{j+1}t_jt_{j-1}$.

Initially we create a trivial triangulation for ∞ and the first two vertices u and v, by introducing three arcs a_1, a_2, a_3 and their mates b_1, b_2, b_3, together with three instruction nodes $\lambda_0, \lambda_1, \lambda_2$ that divide the universe into "triangles" $\Delta \infty uv$ and $\Delta \infty vu$ (actually degenerate wedges $\angle \bar{v}uv$ and $\angle \bar{u}vu$):

$$
\begin{array}{lll}
vert(a_1) = v\,, & next(a_1) = a_2\,, & inst(a_1) = \lambda_1\,; \\
vert(a_2) = \Lambda\,, & next(a_2) = a_3\,, & inst(a_2) = \lambda_1\,; \\
vert(a_3) = u\,, & next(a_3) = a_1\,, & inst(a_3) = \lambda_1\,; \\
vert(b_1) = u\,, & next(b_1) = b_3\,, & inst(b_1) = \lambda_2\,; \\
vert(b_2) = v\,, & next(b_2) = b_1\,, & inst(b_2) = \lambda_2\,; \\
vert(b_3) = \Lambda\,, & next(b_3) = b_2\,, & inst(b_3) = \lambda_2\,;
\end{array}
$$

$$
\lambda_0 = (u, v, \lambda_1, \lambda_2)\,;
$$
$$
\lambda_1 = (\Lambda, a_2, -, -)\,;
$$
$$
\lambda_2 = (\Lambda, b_3, -, -)\,. \tag{18.1}
$$

There is a variable j, initially 3, such that arcs $a_1, \ldots, a_j, b_j, \ldots, b_1$ are in use. The algorithm now proceeds as follows to add a new vertex p to the triangulation:

Step T1. [Follow branch instructions.] Set λ to the root node λ_0. Then if node $\lambda = (\rho_\lambda, q_\lambda, \alpha_\lambda, \beta_\lambda)$ is not a terminal node, set λ to α_λ or β_λ according as $\infty pp_\lambda q_\lambda$ is true or false. Repeat until λ is a terminal node $(\Lambda, a_\lambda, -, -)$.

Step T2. [Subdivide a triangle or wedge.] Set $a \leftarrow a_\lambda$, $b \leftarrow next(a)$, $c \leftarrow next(b)$, $q \leftarrow vert(a)$, $r \leftarrow vert(b)$, $s \leftarrow vert(c)$. (If $q \neq \Lambda$, we have located p within a triangle Δqrs of the existing triangulation; hence $qrsp$ holds, and pq, pr, and ps must be edges of the new triangulation, as remarked above. If $q = \Lambda$, point p lies in the wedge $\angle \bar{r}'rs$, and in particular we have ∞rsp; hence $p\infty$, pr, and ps must become Delaunay edges.) Increase j by 3, thereby making three new arcs a_{j-2}, a_{j-1}, a_j and their mates b_{j-2}, b_{j-1}, b_j available for use. Allocate three new terminal nodes $\lambda' = (\Lambda, a, -, -)$, $\lambda'' = (\Lambda, a_j, -, -)$, $\lambda''' = (\Lambda, c, -, -)$. Set

$$
\begin{array}{lll}
vert(a_j) = q\,, & next(a_j) = b\,, & inst(a_j) = \lambda''\,, \\
vert(a_{j-1}) = r\,, & next(a_{j-1}) = c\,, & inst(a_{j-1}) = \lambda'''\,, \\
vert(a_{j-2}) = s\,, & next(a_{j-2}) = a\,, & inst(a_{j-2}) = \lambda'\,, \\
vert(b_j) = p\,, & next(b_j) = a_{j-2}\,, & inst(b_j) = \lambda'\,, \\
vert(b_{j-1}) = p\,, & next(b_{j-1}) = a_j\,, & inst(b_{j-1}) = \lambda''\,, \\
vert(b_{j-2}) = p\,, & next(b_{j-2}) = a_{j-1}\,, & inst(b_{j-2}) = \lambda'''\,;
\end{array} \tag{18.2}
$$

also change $next(a) \leftarrow b_j$, $inst(a) \leftarrow \lambda'$, $next(b) \leftarrow b_{j-1}$, $inst(b) \leftarrow \lambda''$, $next(c) \leftarrow b_{j-2}$, $inst(c) \leftarrow \lambda'''$. (We have subdivided Δqrs into three triangles Δqps, Δqrp, and Δspr, by changing the arc structure; see Figure 2. We must still change the branching structure so that it will lead to the corresponding terminal nodes λ', λ'', λ''' at appropriate times.) If $q = \Lambda$, go to Step T4, otherwise continue with Step T3.

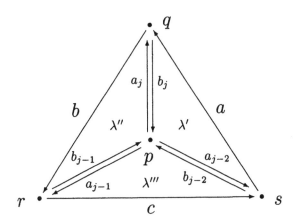

Figure 2. Trisection of a triangle in step T2.

Step T3. [Compile a three-way branch.] Allocate two new branch nodes $\nu = (q, p, \lambda', \lambda'')$ and $\nu' = (s, p, \lambda''', \lambda')$; change node λ to (r, p, ν, ν'). Go to Step T5. (The three instructions λ, ν, ν' can be paraphrased as follows:

"Suppose you're in Δqrs; then
    if left of rp,
        if left of qp, you're in Δqps, **else** you're in Δqrp;
    else if left of sp, you're in Δspr, **else** you're in Δqps."

We have essentially replaced terminal node λ by three terminal nodes λ', λ'', λ'''.)

Step T4. [Compile two-way branches for affected wedges.] Allocate a new branch node $\nu = (s, p, \lambda''', \lambda')$ and change node λ to (r, p, λ'', ν). (At this point we also want to fix up the convex hull, which has just gained point p but it might lose point s and other points clockwise from s.) Set $\mu \leftarrow \nu$, $d \leftarrow next(mate(a))$, $t \leftarrow vert(d)$, and repeat the following operations while $t \neq r$ and ∞pst: Allocate a new terminal node $\nu = (\Lambda, d, -, -)$, change the fourth component β_μ of node μ from λ' to $inst(d)$, set $\mu \leftarrow inst(d)$, change terminal node μ to (t, p, ν, λ'), and perform the subroutine $flip(a, mate(a), d, s, \Lambda, t, p, \nu, \lambda')$ described below; then set $a \leftarrow next(mate(a))$, $d \leftarrow next(mate(a))$, $s \leftarrow t$, $t \leftarrow vert(d)$, and set the second component of terminal node λ' to a. After finishing that loop, allocate another new terminal node $\nu = (\Lambda, next(d), -, -)$; change terminal node $inst(d)$ to (s, p, ν, λ'), and set $inst(x) \leftarrow \nu$ for $x = d$, $next(d)$, and $next(next(d))$. (Translation: Suppose $r = t_j$ and $s = t_{j-1}$ in the current convex hull, and suppose that the point p satisfies $\infty pt_j t_{j+1}$, $\infty pt_j t_{j-1}$, $\infty pt_{j-1} t_{j-2}$, ..., $\infty pt_{k+1} t_k$, and $\infty pt_{k-1} t_k$, where $k \leq j - 1$ and subscripts are treated modulo m. Then point p is being added to the hull, and points $t_{j-1}, \ldots, t_{k+1}$ are being deleted. The new instructions can be paraphrased as follows, assuming that l_i labels the instruction corresponding to former wedge $\angle \bar{t}_{i+1} t_i t_{i-1}$:

l_j: if left of $t_j p$, you're in $\angle \bar{t}_{j+1} t_j p$;
                else if left of $t_{j-1} p$, you're in $\Delta t_j t_{j-1} p$;
                else goto l_{j-1};
l_{j-1}:   if left of $t_{j-2} p$, you're in $\Delta t_{j-1} t_{j-2} p$, else goto l_{j-2};

$\qquad\qquad \vdots$

l_{k+1}:   if left of $t_k p$, you're in $\Delta t_{k+1} t_k p$, else goto l_k;
l_k: if left of $t_k p$, you're in $\angle \bar{p} t_k t_{k-1}$, else you're in $\angle \bar{t}_j p t_k$.

However, 'goto l_k' is actually simplified to 'you're in $\angle \bar{t}_j p t_k$', to avoid a redundant test. We will prove below that these instructions correctly place a new point into a triangle or wedge.) Set $r \leftarrow s$. (This value of r will terminate Step T5 at the appropriate time.)

Step T5. [Find any remaining new triangles.] Set $d \leftarrow mate(c)$, $e \leftarrow next(d)$, $t \leftarrow vert(d)$, $t' \leftarrow vert(c)$, and $t'' \leftarrow vert(e)$. (At this point we know that pt and pt' will be Delaunay edges, and we wish to know whether there ought to be at least one more new Delaunay edge between them. For this we must look at the triangle to the right of arc c, which is the triangle to the left of arc d, namely $\Delta t t'' t'$.) If $t'' \neq \Lambda$ and $t'' t' t p$, allocate new terminal nodes $\nu = (\Lambda, e, -, -)$, $\nu' = (\Lambda, d, -, -)$, change both terminal nodes $inst(c)$ and $inst(d)$ to the branch node (t'', p, ν, ν'), perform the subroutine $flip(c, d, e, t, t', t'', p, \nu, \nu')$ described below, set $c \leftarrow e$, and go back to the beginning of Step T5. Otherwise, if $t' \neq r$, set $c \leftarrow next(mate(next(c)))$ and go back to the beginning of Step T5. Otherwise terminate the updating process. (We have gone all around the cycle surrounding p and returned to vertex r.) $\square$

Steps T4 and T5 use a subroutine $flip(c, d, e, t, t', t'', p, \nu, \nu')$ with the following specifications. We have $d = mate(c)$, $e = next(d)$, $t = vert(d)$, $t' = vert(c)$, $t'' = vert(e)$. The triangles $\Delta t t' p$ and $\Delta t' t t''$ to the left and right of arc c have vertices satisfying the incircle predicate $t'' t' t p$; we want to replace them in the current triangulation by $\Delta p t t''$ and $\Delta t'' t' p$, corresponding to terminal nodes ν and ν'. Thus arc c from t to t' and its mate d from t' to t are being "flipped," i.e., replaced by a pair of arcs between p and t''. The necessary changes to the arc structure are straightforward: Set $e' \leftarrow next(e)$, $c' \leftarrow next(c)$, $c'' \leftarrow next(c')$; then set $next(e) \leftarrow c$, $next(c) \leftarrow c''$, $next(c'') \leftarrow e$, $inst(e) \leftarrow inst(c) \leftarrow inst(c'') \leftarrow \nu$, and $vert(c) \leftarrow p$; also set $next(d) \leftarrow e'$, $next(e') \leftarrow c'$, $next(c') \leftarrow d$, $inst(d) \leftarrow inst(e') \leftarrow inst(c') \leftarrow \nu'$, and $vert(d) \leftarrow t''$. (Notice that arcs c and d are still mates.)

Our description of dag triangulation has been rather lengthy because we have spelled out all of the data structure operations. But in fact the algorithm is shorter and simpler than the similar algorithm sketched in high-level terms in [36]. The main difference is that the algorithm of [36] requires three special points at infinity, two of which can enter simultaneously into incircle tests; the implementation of such an incircle test requires a lengthy program based on a detailed case analysis of limiting behavior, or with special provisions needed to combat degenerate situations. The present algorithm avoids these complications by requiring only a single ideal point ∞, and we will prove that it works in any CCC system. Incidentally, the maintenance of regions outside the convex hull in terms of "wedges" $\angle \bar{t}_{j+1} t_j t_{j-1}$ turns out to be crucial; schemes that allow more general subsets of the halfplanes to the left of $t_j t_{j-1}$, analogous to the simple mechanism of daghull in section 12 above, can fail when several points need to be deleted simultaneously from the current convex hull. Any point lying in such a halfplane will be a new extreme point, but the branching structure will not continue to be correct unless the exterior regions are maintained carefully.

Most of the dag triangulation algorithm is straightforward and easily verified. For example, the instructions compiled in Step T3 subdivide a triangle properly in any CC system; only two cases are not quite immediate:

$$vqr \wedge vrs \wedge vsq \wedge vrp \wedge vqp \implies vps, \tag{18.3}$$

$$vqr \wedge vrs \wedge vsq \wedge vpr \wedge vps \implies vqp. \tag{18.4}$$

These implications are needed to conclude that $v \in \Delta qps$; and they are obviously equivalent to Axioms 5' and 5, respectively.

The instructions compiled during the splitting operation in Step T5 also need to be validated. For this it suffices to show that

$$(v \in \Delta pqr \vee v \in \Delta rqs) \wedge pqr \wedge rqs \wedge vsp \implies v \in \Delta pqs; \tag{18.5}$$

$$(v \in \Delta pqr \vee v \in \Delta rqs) \wedge pqr \wedge rqs \wedge vps \implies v \in \Delta psr. \tag{18.6}$$

(The additional hypothesis $pqrs$ is true when we do the splitting, but we don't need it.) The mapping $(p, q, r, s) \to (s, r, q, p)$ shows that we can assume without loss of generality that $v \in \Delta pqr$; we want to prove

$$vpq \wedge vqr \wedge vrp \wedge rqs \wedge vsp \implies vqs; \tag{18.7}$$

$$vpq \wedge vqr \wedge vrp \wedge rqs \wedge vps \implies vsr. \tag{18.8}$$

Both of these are easy. If $vrq \wedge vqr \wedge vrp$, then $vsp \wedge vsq \implies vrs$ and $vps \wedge vrs \implies vsq$ by Axioms 5 and 5'; and $vsq \wedge vqr \wedge vrs \implies rsq$ by Axiom 4.

The proof becomes more intricate when we try to validate the instructions compiled in Step T4. Suppose we have $pt_j t_{j+1}, pt_j t_{j-1}, pt_j t_{j-1} t_{j-2}, \ldots, pt_{k+1} t_k, pt_{k-1} t_k$, and $k \le j - 1$, as in the comments on that step. We will prove that every subsequent vertex v that comes through the newly compiled branch instructions will lie in the triangle or wedge claimed.

If v comes to label l_j, we have $v \in \angle \bar{t}_{j+1} t_j t_{j-1}$, which means that $vt_j t_{j+1} \wedge vt_j t_{j-1}$. Therefore if $vt_j p$ we have $v \in \angle \bar{t}_{j+1} t_j p$ by definition. Otherwise if $vpt_j \wedge vt_{j-1} p$ we have $v \in \Delta t_j t_{j-1} p$, again by definition. Otherwise if $vpt_j \wedge vpt_{j-1}$, we have $v \in \angle \bar{t}_j pt_{j-1}$, and the instructions continue at label l_{j-1}.

Now suppose $j - 1 \ge i > k$ and v comes to label l_i. There are two ways this can happen. First, we might have $v \in \angle \bar{t}_{i+1} t_i t_{i-1}$; i.e., $vt_i t_{i+1}$ and $vt_i t_{i-1}$. Then we must have vpt_i; otherwise there would be a vortex from $t_{i+1} \to t_{i-1} \to p \to t_{i+1}$ to v in the tournament for t_i, because $t_i t_{i+1} t_{i-1}$ by definition of convex hull. We must also have $vt_i t_j$; otherwise $j > i + 1$ and that tournament would contain the vortex $v \to t_j \to t_{i-1} \to v$ out of t_{i+1}. Hence $t_i \in \Delta vpt_j$, and we have vpt_j by Axiom 4. It follows that

$$\text{if } vt_{i-1}p \text{ then } v \in \Delta t_i t_{i-1} p \text{ else } v \in \angle \bar{t}_j pt_{j-1} . \tag{18.9}$$

The other way we can get to l_i is from the program for l_{i+1}. In this case, we will see by induction on $j - i$ that we must have $v \in \angle \bar{t}_j pt_i$ and also $vt_{i+1}t_i$. Therefore we must have $vt_i t_{i-1}$, to avoid a vortex $t_{i+1} \to t_{i-1} \to p \to t_{i+1}$ out of v in the tournament for t_i. And therefore (18.9) holds in this case as well.

Finally, if v comes to label l_k, we have $v \in \angle \bar{t}_{k+1} t_k t_{k-1}$. We must now have $vt_k t_j$; this is obvious if $j = k + 1$, and otherwise the tournament for t_k would contain $v \to t_j \to t_{k-1} \to v$ out of t_{k+1}. It follows that the instruction 'if $vt_k p$ then $v \in \angle \bar{p} t_k t_{k-1}$ else $v \in \angle \bar{t}_j pt_k$' is correct; for if vpt_k we have $t_k \in \Delta vpt_j$, and vpt_j must hold by Axiom 4.

This completes the proof that the algorithm's branching structure correctly locates points in triangles or wedges as claimed. The remaining logic, concerning the incircle test in Step T5, is justified by the remarks about incremental Delaunay triangulation that we stated before describing the algorithm.

How fast is dag triangulation? Like the daghull algorithm of section 12, it has a worst-case running time of order N^2 if we present it with an N-gon whose vertices are input in cyclic order. But its expected behavior on randomized inputs is quite reasonable:

Theorem. *If the dag triangulation algorithm is applied to N points of any CCC system in random order, it will on the average compute $O(N)$ instructions and make $O(N \log N)$ calls on the incircle predicate, of which all but $O(N)$ are simple counterclockwise tests having the special form ∞pqr.*

Proof. First let's consider the total number of nodes. There are three after initialization (18.1), and we add three more each time we perform Step T2. Step T3 adds two whenever a triangle is deleted from the current triangulation. (A "triangle" in this proof is a finite triangle Δpqr, not a wedge.) Triangles also leave the current triangulation whenever Step T3 is performed; this occurs a total of $N - M - C$ times, where M is the number of points in the final convex hull and C is the number of points that were in the current convex hull at one time but not in the final convex hull. Therefore the total number of nodes is $5N - 7 + C + 2(T - (N - M - C)) = 3N + 3C + 2M + 2T - 7$, where T is the number of triangles that were in the current triangulation at one time but not in the final triangulation. We will prove below that T is $O(N)$ on the average.

Notice that $T - (N - M - C)$ is also the number of incircle tests in Step T5 that turn out to be true. The number of incircle tests that turn out to be false in that step is always equal to the number that turn out to be true, plus 3 if we came to T5 through T3, or plus $1 + c$ if we came through T4 with c points deleted from the hull. (Actually this is a slight overcounting: Step T5 does not test $t''t'tp$ when $t'' = \Lambda$, because the predicate $\infty t'tp$ is known to be false in that case. So some of the "false" incircle tests counted here are not actually performed.) We conclude that the total number of incircle tests on finite points $t''t'tp$ is at most $2(T - (N - M - C)) + 3(N - M - C) + M + C - 3 + C = 2T + N + C - 3$.

The total execution time for Steps T2–T5 is therefore $O(N + T)$; the "inner loop" must be the counterclockwise tests $\infty pp_\lambda q_\lambda$ made in Step T1. Those tests are of the following kinds, depending on when node λ became a branch node: The initial node λ_0 of (18.1) clearly leads to $N - 2$ tests and we can ignore it. The three branch nodes λ, ν, ν' of Step T3 lead to instructions that are performed for all subsequent points that lie inside a triangle Δqrs that is leaving the triangulation. Step T4 is a bit more complicated and we will return to it in a moment. The branch node $inst(d)$ of Step T5 leads to an instruction that is performed for all subsequent points in the triangle $\Delta tt''t'$ leaving the triangulation; and the branch node $inst(c)$ of that step leads to an instruction that is performed for all subsequent points in $\Delta ptt'$. Triangle $\Delta ptt'$ has not necessarily been part of any triangulation, but any point $v \in \Delta ptt'$ does satisfy the incircle predicate $tt''t'v$. (Proof: By (17.5) we have $v \in \Delta ptt' \implies ptt'v$; i.e., $tt'pv$. We also have $tt'\infty v$, $tt'\infty p$, $tt'pt''$, and $tt't''\infty$. So Axiom 5' would fail in the CC system fixing t if we had $tt't''v$.) Therefore we can "charge" the branch instruction $inst(c)$ to $\Delta tt''t'$, in the following argument.

The remainder of the proof is essentially identical to the analysis of a similar algorithm in [36], so it will only be sketched here; complete details can be found in that paper. Let us say that a triple of finite points qrs with counterclockwise orientation ∞qrs has *scope* k if there are k other points v such that $qrsv$ holds. Then Δqrs is part of the Delaunay triangulation iff qrs has scope 0, and the triples with large scope are unlikely to appear in the partial triangulations. The argument in the previous paragraph shows that a triple with scope k will be charged for the execution of at most $2k$ instructions in Step T1, and only if its points $\{q, r, s\}$ are entered before the other k points in its scope. The latter event occurs with probability $3!\,k!/(k + 3)! = 6/(k + 1)(k + 2)(k + 3)$.

If there are T_k triples of scope k, let $T_{<k} = T_0 + T_1 + \cdots + T_{k-1}$. The expected number of triangles T that enter the triangulation and leave it again satisfies

$$T + 2N - M - 2 = \sum_{k \geq 0} \frac{6T_k}{(k + 1)(k + 2)(k + 3)}$$

$$= \sum_{k \geq 1} \frac{18T_{<k}}{k(k + 1)(k + 2)(k + 3)}, \tag{18.10}$$

because there are $2N - M - 2$ triangles (and M wedges) in the final triangulation on N finite points. Similarly, the expected number of branch instructions executed in T1 after being generated in T3 and T5 is at most $2B$, where

$$B = \sum_{k \geq 0} \frac{6k\,T_k}{(k + 1)(k + 2)(k + 3)} = \sum_{k \geq 1} \frac{(12k - 18)\,T_{<k}}{k(k + 1)(k + 2)(k + 3)}. \tag{18.11}$$

It will follow that $T = O(N)$ and $B = O(N \log N)$ if we can prove the upper bound

$$T_{<k} = O(k^2 N). \tag{18.12}$$

(Note that the tails of the sums for $k \geq N$ in (18.10) and (18.11) are respectively $O(1)$ and $O(N)$, because $T_{<k} = \binom{N}{3}$ when $k \geq N$.)

The Clarkson-Shor probabilistic method [12] can be used to establish (18.12) as follows: The expected number E_r of triangles in the triangulation of r randomly chosen points satisfies

$$2r > E_r = \sum_{j \geq 0} \frac{\binom{N-j-3}{r-3}}{\binom{N}{r}} T_j \geq \frac{\binom{N-k-2}{r-3}}{\binom{N}{r}} T_{<k},$$

when $k \leq N - 2$. Choosing $r = \lfloor 2N/(k+1) \rfloor + 1$ gives the desired bound, after some manipulation.

We still need to consider the branch instructions compiled in Step T4. We've seen that the instructions labeled l_i for $j \geq i \geq k$ are performed only for vertices v that satisfy $\infty v t_i t_{i-1}$; so we can charge the execution of those instructions to the edge $t_{i-1} t_i$ that was formerly in the convex hull. The total cost is $O(N \log N)$, by the theorem in section 12 above. $\square$

Computational experiments with the same data used to test convex hull algorithms in section 13 shows that dag triangulation makes about three to fifteen times as many memory references as daghull, depending on the type of input data. Here are the actual statistics:

data	daghull mems+ccs	dag triangulation mems+ccs+incircles
128 cities	3543+1034	29853+2265+939
100 uniform	2028+579	20260+1219+671
1000 uniform	21078+6811	263344+22666+8659
10000 uniform	210795+69806	2994832+337812+89233
100 n-gon	4530+1079	17883+1311+343
1000 n-gon	63204+16737	285263+21547+5530
1000 n-gon	859311+243106	3308159+309298+59722
100 nested	7544+2174	25045+3026+500
1000 nested	151002+48855	455438+78955+8785
10000 nested	2318715+769733	6135645+1349280+91992

(The CCC system used in these n-gon tests was the degenerate case when all coordinates are equal, as explained in section 19 below; points were inserted in random order of their serial numbers. The final triangulation then contains all edges of the forms $a_0 - a_k$, $a_k - a_{k+1}$, and $a_k - \infty$, where a_k is the point with serial number k.) The program for dag triangulation was 167 statements long, not including the code for the cc or incircle procedures, but including the procedure that allocates space for a new node. (This compares to 55 statements for daghull, the simplest of the randomized $O(N \log N)$ algorithms for convex hull.)

The dag triangulation algorithm can be run without any artificial infinite point by letting any real point play the role of ∞. The resulting triangulation will be the same as with artificial ∞, except that there will be additional edges joining certain vertices of the convex hull. These additional edges, together with the convex hull itself, define a triangulation that is dual to the Voronoi diagram for furthest points instead of closest points. Of course the algorithm runs more slowly when it has to make full incircle tests each time instead of simpler counterclockwise tests most of the time; but this observation can be useful when debugging. A sample run on the 128-city problem without an artificial ∞ made 30199 memory references and 3214 incircle tests. The additional edges it found between cities of the convex hull ran from St. Johnsbury to Regina, Vancouver, and West Palm Beach; from Vancouver also to Salem, Santa Rosa, San Francisco, and West Palm Beach; and from West Palm Beach also to Salinas, Santa Barbara, and San Diego.

19. Incircle degeneracy

Let us now return to the ideas of section 14, where we developed methods to define CC systems on arbitrary sequences of points in the plane, allowing two points to be coincident and three points to be collinear. Now we want to extend those methods, so that CCC systems can be defined when we also allow four points to be cocircular.

From a practical standpoint, our best rule in section 14 was derived from the small perturbations defined in (14.16). The same perturbations also give us a satisfactory way to define $pqrs$ when the determinant $|pqrs|$ is zero. Suppose $p_1 \prec p_2 \prec \cdots \prec p_n$ is any linear ordering of the points, not necessarily related to lexicographic order, and define $p_1', \ldots, p_n'$ by (14.16). We will say that $pqrs$ is true iff the first nonvanishing coefficient of the determinant $|p'q'r's'|$ is positive when that determinant is expanded in increasing powers of ϵ. We also say that ∞pqr is true iff the first nonvanishing coefficient of $|p'q'r'|$ is positive; this gives the rule determined by the algorithm at the end of section 14.

If $p \prec q \prec r \prec s$, we have

$$
\begin{aligned}
|p'q'r's'| = {} & |pqrs| + \epsilon_s \, f(p,q,r,s) + \delta_s \, g(p,q,r,s) \\
& - \epsilon_r \, f(s,p,q,r) - \delta_r \, g(s,p,q,r) \\
& + \epsilon_q \, f(r,s,p,q) + \delta_q \, g(r,s,p,q) \\
& - \epsilon_p \, f(q,r,s,p) - \delta_p \, g(q,r,s,p) + O(\epsilon_s^2) \,,
\end{aligned} \tag{19.1}
$$

where the functions f and g are defined by

$$
|pqrs'| = |pqrs| + \epsilon_s \, f(p,q,r,s) + \delta_s \, g(p,q,r,s) + O(\epsilon_s^2) \,. \tag{19.2}
$$

Let p^s and Δ_{ps}^2 be defined by (17.2). An extension of the trick by which we showed earlier that $|spqr|$ has the same sign as $|p^s q^s r^s|$ can be used to find simple formulas for $f(p,q,r,s)$ and $g(p,q,r,s)$: We have

$$
\det \begin{pmatrix} x_p & y_p & x_p^2 + y_p^2 & 1 \\ x_q & y_q & x_q^2 + y_q^2 & 1 \\ x_r & y_r & x_r^2 + y_r^2 & 1 \\ x_s - \delta & y_s + \epsilon & (x_s - \delta)^2 + (y_s + \epsilon)^2 & 1 \end{pmatrix} = \det \begin{pmatrix} x_p - x_s & y_p - y_s & \Delta_{ps}^2 & 1 \\ x_q - x_s & y_q - y_s & \Delta_{qs}^2 & 1 \\ x_r - x_s & y_r - y_s & \Delta_{rs}^2 & 1 \\ -\delta & \epsilon & \delta^2 + \epsilon^2 & 1 \end{pmatrix};
$$
$$\tag{19.3}$$

hence

$$f(p, q, r, s) = \det \begin{pmatrix} x_p - x_s & \Delta^2_{ps} & 1 \\ x_q - x_s & \Delta^2_{qs} & 1 \\ x_r - x_s & \Delta^2_{rs} & 1 \end{pmatrix}, \tag{19.4}$$

$$g(p, q, r, s) = \det \begin{pmatrix} y_p - y_s & \Delta^2_{ps} & 1 \\ y_q - y_s & \Delta^2_{qs} & 1 \\ y_r - y_s & \Delta^2_{rs} & 1 \end{pmatrix}. \tag{19.5}$$

Lemma. If $|pqrs| = 0$ and if p, q, r are distinct, then either $f(p, q, r, s) \neq 0$ or $g(p, q, r, s) \neq 0$.

Proof. Without loss of generality, we can assume that $s = (0, 0)$. Suppose first that $p = (0, 0)$ and that $f(p, q, r, s) = g(p, q, r, s) = 0$; this means

$$\det \begin{pmatrix} x_q & x_q^2 + y_q^2 \\ x_r & x_r^2 + y_r^2 \end{pmatrix} = \det \begin{pmatrix} y_q & x_q^2 + y_q^2 \\ y_r & x_r^2 + y_r^2 \end{pmatrix} = 0,$$

while $x_q^2 + y_q^2$ and $x_r^2 + y_r^2$ are nonzero. So

$$\frac{x_q}{x_q^2 + y_q^2} = \frac{x_r}{x_r^2 + y_r^2}, \qquad \frac{y_q}{x_q^2 + y_q^2} = \frac{y_r}{x_r^2 + y_r^2}, \qquad \frac{x_q^2 + y_q^2}{(x_q^2 + y_q^2)^2} = \frac{x_r^2 + y_r^2}{(x_r^2 + y_r^2)^2},$$

and we must have $x_q^2 + y_q^2 = x_r^2 + y_r^2$; hence $x_q = x_r$ and $y_q = y_r$, a contradiction.

If $s = (0, 0)$ and none of p, q, r is $(0, 0)$, we have $|pqrs| = 0$ iff the points $1/z_p$, $1/z_q$, and $1/z_r$ are collinear in the complex plane, where $z_p = x_p + iy_p$. Let $1/z_p = u_p + iv_p = (x_p - iy_p)/(x_p^2 + y_p^2)$, and define $u_q, v_q, u_r,$ and v_r similarly. Then

$$f(p, q, r, 0) = |z_p|^2 |z_q|^2 |z_r|^2 \det \begin{pmatrix} u_p & 1 & u_p^2 + v_p^2 \\ u_q & 1 & u_q^2 + v_q^2 \\ u_r & 1 & u_r^2 + v_r^2 \end{pmatrix}, \tag{19.6}$$

$$g(p, q, r, 0) = |z_p|^2 |z_q|^2 |z_r|^2 \det \begin{pmatrix} -v_p & 1 & u_p^2 + v_p^2 \\ -v_q & 1 & u_q^2 + v_q^2 \\ -v_r & 1 & u_r^2 + v_r^2 \end{pmatrix}. \tag{19.7}$$

So the lemma boils down to proving that we cannot have

$$\det \begin{pmatrix} u_p & u_p^2 + v_p^2 & 1 \\ u_q & u_q^2 + v_q^2 & 1 \\ u_r & u_r^2 + v_r^2 & 1 \end{pmatrix} = \det \begin{pmatrix} v_p & u_p^2 + v_p^2 & 1 \\ v_q & u_q^2 + v_q^2 & 1 \\ v_r & u_r^2 + v_r^2 & 1 \end{pmatrix} = 0 \tag{19.8}$$

when (u_p, v_p), (u_q, v_q), and (u_r, v_r) are distinct, collinear points.

We have $(u_r, v_r) - (u_p, v_p) = \lambda((u_q, v_q) - (u_p, v_p))$ for some $\lambda \neq 0, 1$. Consequently, letting $\Delta^2 = (u_q - u_p)^2 + (v_q - v_p)^2$, we have

$$\det \begin{pmatrix} u_p & u_p^2 + v_p^2 & 1 \\ u_q & u_q^2 + v_q^2 & 1 \\ u_r & u_r^2 + v_r^2 & 1 \end{pmatrix} = \det \begin{pmatrix} 0 & v_p^2 & 1 \\ u_q - u_p & (u_q - u_p)^2 + v_q^2 & 1 \\ u_r - u_p & (u_r - u_p)^2 + v_r^2 & 1 \end{pmatrix}$$

$$= (u_q - u_p) \det \begin{pmatrix} 0 & v_p^2 & 1 \\ 1 & v_p^2 + 2v_p(v_q - v_p) + \Delta^2 & 1 \\ \lambda & v_p^2 + 2\lambda v_p(v_q - v_p) + \lambda^2 \Delta^2 & 1 \end{pmatrix}$$

$$= (u_q - u_p) \det \begin{pmatrix} 0 & 0 & 1 \\ 1 & 1 & 1 \\ \lambda & \lambda^2 & 1 \end{pmatrix} = (u_q - u_p)(\lambda^2 - \lambda) \Delta^2.$$

Consequently the first determinant of (19.8) is zero iff $u_p = u_q = u_r$, and the second is zero iff $v_p = v_q = v_r$. □

The lemma tells us that the first seven terms of (19.1) will define $pqrs$ whenever the set $\{p, q, r, s\}$ contains at least three distinct points. Conversely, it is easy to see that $f(p, q, r, s) = g(p, q, r, s) = 0$ whenever p, q, r are not distinct; hence the first nine terms of (19.1) are zero whenever $\{p, q, r, s\}$ contains at most two distinct points. Equation (19.3) shows that the coefficients of ϵ_s^2, δ_s^2, ϵ_r^2, δ_r^2, ϵ_q^2, δ_q^2, ϵ_p^2, and δ_p^2 will also be zero in that case.

To break ties in cases of extreme degeneracy, the first possibly nonvanishing coefficient will therefore be the coefficient of $\epsilon_s \epsilon_r$, which is the coefficient of ϵ_r in $f(p, q, r', s)$, namely $2(y_r - y_s)(x_q - x_p)$. If this too is zero, we turn to the coefficient of $\epsilon_s \delta_r$, which is $(x_q - x_r)^2 + (y_q - y_s)^2 - (x_p - x_r)^2 - (y_p - y_s)^2$. One of these two is bound to be nonzero unless $p = q$. For if $p \neq q$ there are two cases: Either $r = s$, in which case we must have $r = p$ or $r = q$, and the coefficient of $\epsilon_s \delta_r$ is $\pm \Delta_{pq}^2$; or $r \neq s$ and we must have $\{r, s\} = \{p, q\}$. In the latter case the coefficient of $\epsilon_s \delta_r$ can be zero only if $(x_q - x_p)^2 = (y_q - y_p)^2$; and then $x_q - x_p$ and $y_q - y_p$ must both be nonzero, so the coefficient of $\epsilon_s \epsilon_r$ will not vanish.

If the coefficients of $\epsilon_s \epsilon_r$ and $\epsilon_s \delta_r$ are zero, we try the (similar) coefficients of $\epsilon_s \epsilon_q$ and $\epsilon_s \delta_q$. Those will both turn out to be zero only if $p = q = r$, in which case the coefficients of $\epsilon_s \epsilon_p$ and $\epsilon_s \delta_p$ will vanish too, as will the coefficient of $\epsilon_r \epsilon_q$. We will, however, find a term in $\epsilon_r \delta_q$ if $r \neq s$.

Finally, if $p \prec q \prec r \prec s$ but $p = q = r = s$, we can assume without loss of generality that $x_p = y_p = \cdots = x_s = y_s = 0$. Now it is clear that the first nonzero term of $|p'q'r's'|$ is $\epsilon_s^2 \epsilon_r \delta_q$; we therefore consider $pqrs$ to be true, in this maximally degenerate case.

Our rule for defining $pqrs$ in general, given arbitrary points $p = (x_p, y_p)$, ..., $s = (x_s, y_s)$ in the plane, therefore boils down to the following. As in section 14, we attach a unique serial number l_p to each point p.

Step 1.   Evaluate the determinant

$$|pqrs| = \det \begin{pmatrix} x_p - x_s & y_p - y_s & \Delta_{ps}^2 \\ x_q - x_s & y_q - y_s & \Delta_{qs}^2 \\ x_r - x_s & y_r - y_s & \Delta_{rs}^2 \end{pmatrix} \qquad (19.9)$$

with perfect accuracy, where $\Delta_{ps}^2 = (x_p - x_s)^2 + (y_p - y_s)^2$. If the result is nonzero, return 'true' if it is positive, 'false' if it is negative. Otherwise set $b = $ 'true' and proceed to Step 2.

Step 2.   If $l_p > l_q$, interchange $p \leftrightarrow q$ and complement the value of b; if $l_q > l_r$, interchange $q \leftrightarrow r$ and complement the value of b; if $l_r > l_s$, interchange $r \leftrightarrow s$ and complement the value of b; repeat until $l_p < l_q < l_r < l_s$.

Step 3.   Compute the following quantities exactly, until finding the first nonzero result, then complement b if that result is negative:

$$\begin{array}{c} f(p,q,r,s)\,, \ g(p,q,r,s)\,, \ f(q,p,s,r)\,, \ g(q,p,s,r)\,, \\ f(r,s,p,q)\,, \ g(r,s,p,q)\,, \ h(p,q,r,s)\,, \ j(p,q,r,s)\,, \\ h(r,p,q,s)\,, \ j(r,p,q,s)\,, \ j(p,s,q,r)\,, \ +1\,. \end{array} \qquad (19.10)$$

Here

$$f(p,q,r,s) = \det \begin{pmatrix} x_p - x_r & \Delta_{ps}^2 - \Delta_{rs}^2 \\ x_q - x_r & \Delta_{qs}^2 - \Delta_{rs}^2 \end{pmatrix}, \qquad (19.11)$$

$$g(p,q,r,s) = \det \begin{pmatrix} y_p - y_r & \Delta_{ps}^2 - \Delta_{rs}^2 \\ y_q - y_r & \Delta_{qs}^2 - \Delta_{rs}^2 \end{pmatrix}, \qquad (19.12)$$

$$h(p,q,r,s) = (x_q - x_p)(y_r - y_s)\,, \qquad (19.13)$$

$$j(p,q,r,s) = (x_q - x_r)^2 + (y_q - y_s)^2 - (x_p - x_r)^2 - (y_p - y_s)^2\,. \qquad (19.14)$$

Step 4.   Return the value of b.

Examples exist in which each of the 12 quantities in (19.10) will be the first nonzero number of the sequence.

As in section 14, we can use this approach to obtain a highly robust algorithm for Delaunay triangulation, producing a unique answer (once the serial numbers are assigned) that agrees with data that has been perturbed at most a small percentage of the total range. However, in this case the conversion to fixed-point arithmetic must use the same scale factor in both dimensions: The x and y coordinates should be rounded respectively to the nearest values of the form $x/2^d$ and $y/2^d$, where x and y are integers in the ranges $x_0 \le x_0 + 2^{b_x}$ and $y_0 \le y < y_0 + 2^{b_y}$ and where all input data lies in the rectangle with corners $(x_0, y_0)/2^d$ and $(x_0 + 2^{b_x}, y_0 + 2^{b_y})/2^d$. Then we need to do exact arithmetic on integers with $3 \max(b_x, b_y) + \min(b_x, b_y) + 3$ bits of precision and a sign bit; with floating point arithmetic it is also possible to get the correct sign of the determinant with one less bit of precision. Thus, for example, we can go up to $b_x = 13$, $b_y = 12$, if we evaluate (19.9) with IEEE standard double precision arithmetic. We found earlier that 26-bit input data could be handled by the IEEE standard when we were simply finding convex hulls; we need about twice as many bits to compute $|pqrs|$ as to compute $|pqr|$.

Fixed-point arithmetic is not the only available option. We can also compute robust Delaunay triangulations with a suitable floating-point scheme: Suppose each x and y coordinate is rounded to the nearest value that is either 0 or has the form

$$\pm(1 + m2^{-b})2^k \, , \text{ where } 0 \le m < 2^b \text{ and } |k| < K \, . \tag{19.15}$$

Then the determinant $|pqrs|$ will be the sum of at most 12 numbers of the form $\pm(1 + m2^{-4b-3})2^k$, where $0 \le m < 2^{4b+3}$ and $|k| < 4K$. The sum of such numbers is not generally representable in the same form, but we can readily determine the sign of such a sum with no loss of accuracy. This form of representation would be appropriate when calculating the Delaunay triangulation for points (ξ, η, ζ) on the unit sphere, possibly defined for latitude θ and longitude ϕ by the formulas

$$\xi = \sin\phi\cos\theta \, , \ \eta = \cos\phi\cos\theta \, , \ \zeta = \sin\theta \, . \tag{19.16}$$

Projecting this point onto the plane via (17.8), when $\theta \ne \pi/2$, gives

$$x = \sin\phi \cot\left(\frac{\pi}{4} - \frac{\theta}{2}\right) \, , \quad y = \cos\theta \cot\left(\frac{\pi}{4} - \frac{\theta}{2}\right) \, ; \tag{19.17}$$

floating-point approximations near these true values can then be found, having errors that correspond to small changes in θ and ϕ. (Note that the "obvious" approach, in which ξ, η, and ζ are rounded to points that are "nearly" on the sphere, is not valid; it does not lead to determinants whose signs obey Axioms C1–C5.)

20. Generalization to higher dimensions

A *hypertournament of rank r* is an r-ary predicate defined on all ordered r-tuples of distinct points, with the property that interchanging any two points complements the relation. Thus, a set of triples satisfying Axioms 1–3 of section 1 is a hypertournament of rank 3; a set of quadruples satisfying Axioms C1–C3 of section 17 is a hypertournament of rank 4; and an ordinary tournament is a hypertournament of rank 2.

There are $2^{\binom{n}{r}}$ ways to define a hypertournament of rank r on n labelled points, because we can independently choose truth values for a particular ordering of each r-element subset. The *transitive hypertournament* of rank r on the points $\{1, \ldots, n\}$ is defined by the condition that $p_1 \ldots p_r$ is true whenever $1 \le p_1 < \cdots < p_r \le n$; thus, in general, $p_1 \ldots p_r$ is true in the transitive hypertournament if and only if the number of pairs of indices $i < j$ with $p_i > p_j$ is even. The transitive hypertournament of rank 3 on n points is the CC system corresponding to an n-gon. The transitive hypertournament of rank 4 on n points is the CCC system corresponding to n coincident points in the plane, under our rule for eliminating degeneracy in section 19.

Given a hypertournament of rank $r \geq 1$ on n points, the hypertournament of rank $r - 1$ on $n - 1$ points formed by *fixing point p* is obtained by saying that $q_1 \ldots q_{r-1}$ is true in the latter iff $pq_1 \ldots q_{r-1}$ is true in the former. (We have already used this idea to associate tournaments on $n - 1$ points with every point of a CC system.) A sequence of k points $p_1, \ldots, p_k$ can also be fixed, when $r \geq k$, thereby obtaining a hypertournament of rank $r - k$ on $n - k$ points.

Every hypertournament H of rank r on n labelled points, where those points are subject to a linear ordering $a_1 < a_2 < \cdots < a_n$, has a *dual hypertournament H^** of rank $n - r$ on those same points, defined as follows: Let $\{p_1, \ldots, p_r\} \cup \{q_1, \ldots, q_{n-r}\} = \{a_1, \ldots, a_n\}$ be a partition of the points, where $p_1 < \cdots < p_r$ and $q_1 < \cdots < q_{n-r}$. Then $p_1 \ldots p_r$ is true in H iff $q_1 \ldots q_{n-r}$ is true in H^*. For example, suppose the points are $\{1, 2, 3, 4, 5\}$ with the natural ordering. Then the two CC systems in which we have $4 \in \Delta 123$ and $5 \in \Delta 124$ have the triples

$$123, \ 124, \ 125, \ \neg 134, \ \neg 135, \ \neg 145, \ 234, \ 235, \ 245, \ 345 \text{ or } \neg 345, \tag{20.1}$$

and their duals are the tournaments

$$45, \ 35, \ 34, \ \neg 25, \ \neg 24, \ \neg 23, \ 15, \ 14, \ 13, \ 12 \text{ or } \neg 12. \tag{20.2}$$

Negating a point p has the effect of complementing every r-tuple containing p in a hypertournament. The *complement* of a hypertournament is obtained by complementing the value of every r-tuple. If r is odd, we obtain the complement of a hypertournament by negating all its points; if r is even, however, we cannot in general obtain a hypertournament isomorphic to the complementary hypertournament by negating points. For example, the pairs

$$12, \ 31, \ 41, \ 51, \ 16, \ 71, \ 18, \ 23, \ 24, \ 25, \ 62, \ 72, \ 34, \ 53, \ 63, \ 73,$$
$$38, \ 45, \ 64, \ 74, \ 84, \ 56, \ 75, \ 85, \ 67, \ 86, \ 87 \tag{20.3}$$

define a tournament on 8 elements that is not taken into its complement by any signed permutation.

Two hypertournaments H and H' are called *preisomorphic* if there is a signed bijection $p \mapsto p'$ between their points such that we have either $p_1 \ldots p_r$ in $H \Leftrightarrow p'_1 \ldots p'_r$ in H' or $p_1 \ldots p_r$ in $H \Leftrightarrow \neg p'_1 \ldots p'_r$ in H'. In section 5 above, we called CC systems preisomorphic iff there was a signed bijection with $p_1 p_2 p_3 \Leftrightarrow p'_1 p'_2 p'_3$; this simplified definition can be used whenever r is odd, but example (20.3) shows that it cannot be used when $r = 2$. If a_n and b_n denote the number of equivalence classes of ordinary tournaments under signed bijection and under preisomorphism, respectively, we have the following values for small n:

$n =$	1	2	3	4	5	6	7	8	
$a_n =$	1	1	1	2	2	6	17	79	
$b_n =$	1	1	1	2	2	6	17	69.	(20.4)

The dual of a hypertournament, as we have defined it, depends on the linear ordering $a_1 < a_2 < \cdots < a_n$. But all duals obtained from different linear orderings are preisomorphic to each other. For if we decide to interchange, say, the relative order of a_k and a_{k+1}, the $(n - r)$-tuples $q_1 \ldots q_{n-r}$ in the new dual agree with those in the former one except when a_k and a_{k+1} are both present or both absent. Thus we obtain the same effect by negating a_k and a_{k+1}, then complementing the entire dual hypertournament. All orderings can be obtained by repeatedly interchanging adjacent elements. This argument shows, in fact, that all duals obtained from different orderings are obtainable from each other by negating an even number of points and possibly complementing everything; no signed permutations other than "signed identity permutations" are needed.

Negating a point of a hypertournament negates every tuple *not* containing that point, in the dual hypertournament; the same effect is achieved in the dual by negating the point and then complementing everything. Hence, if H is preisomorphic to H', and if H^* and H'^* are duals respectively of H and H', then H^* is preisomorphic to H'^*.

Let us say that a hypertournament of rank r on n points is *geometric* if each tournament obtained by fixing $r - 2$ of its points is vortex-free. Thus, every hypertournament of rank 1 is trivially geometric; a geometric hypertournament of rank 2 is a vortex-free tournament; a geometric hypertournament of rank 3 is a pre-CC system. Any hypertournament that is preisomorphic to a geometric hypertournament is itself geometric.

Lemma. *The dual of a geometric hypertournament is geometric.*

Proof. Let H be a hypertournament of rank r on n points, and let H^* be its dual. If H^* is not geometric, there are points $p_1, \ldots, p_{n-r-2}$ and q_1, q_2, q_3, q_4 such that the tournament obtained from H^* by fixing $p_1 \ldots p_{n-r-2}$ contains a vortex V on q_1, q_2, q_3, q_4. Let the other $r - 2$ points be $p'_1, \ldots, p'_{r-2}$, and let H' be H restricted to the points $\{p'_1, \ldots, p'_{r-2}, q_1, q_2, q_3, q_4\}$. Then the tournament obtained from H' by fixing $p'_1, \ldots p'_{r-2}$ is the dual of V.

But the dual of a vortex is a vortex. For example, if V is a vortex from $q_1 \to q_2 \to q_3 \to q_1$ to v_4, then the true pairs of V are $q_1q_2, q_2q_3, q_3q_1, q_1q_4, q_2q_4, q_3q_4$ and $q_1 < q_2 < q_3 < q_4$; and the true pairs of V^* are $q_3q_4, q_1q_4, q_4q_2, q_2q_3, q_1q_3, q_1q_2$, defining a vortex from q_1 into $q_2 \to q_3 \to q_4 \to q_2$. Preisomorphism also takes vortices into vortices. Therefore H' is not geometric, and neither is H. □

Corollary. *Every geometric hypertournament of rank r on $r + 2$ points is preisomorphic to the transitive hypertournament of rank r on those points.*

Proof. We know from the lemma in section 4 that every vortex-free tournament is preisomorphic to a transitive tournament. Take the dual of that statement. □

We have defined a CCC system to be a hypertournament of rank 4 such that fixing any point yields a CC system. Thus, a CCC system is not only geometric, its associated rank-3 hypertournaments also satisfy Axiom 4. Let's pursue this idea and define a *CCCC system* to be a hypertournament of rank 5 such that fixing any point yields a CCC system.

It turns out that transitive hypertournaments of rank 5 are CCCC systems. Suppose, for example, that $n = 9$, and consider the quadruples that we get by fixing a point, say 3. The resulting hypertournament of rank 4 is precisely what we get from the transitive hypertournament on $\{1, 2, 4, 5, 6, 7, 8, 9\}$ by negating points 1 and 2; this follows because, for example, $13458 = 3\bar{1}458$. If we now fix another point, say 7, we obtain the hypertournament of rank 3 that results when points $\bar{1}$, $\bar{2}$, 4, 5, and 6 are negated in the transitive CC system on $\{\bar{1}, \bar{2}, 4, 5, 6, 8, 9\}$; in other words, it is the result of negating 4, 5, and 6 in the transitive system on $\{1, 2, 4, 5, 6, 8, 9\}$. This is a CC system, because we observed in section 5 that consecutive points of an n-gon can be negated without violating Axiom 4.

We might suppose that CCCC systems correspond somehow to the signs of 5×5 determinants whose rows are something like

$$ x_p \qquad y_p \qquad x_p^2 + y_p^2 \qquad (x_p^2 + y_p^2)^2 \qquad 1\,; $$

but the manipulations that worked in section 17 above do not generalize sufficiently.

In fact, CCCC systems are "the end of the line." If we attempt to define CC-CCC systems as hypertournaments with the property that fixing any point yields a CCCC system, we soon find that there is no such thing as a CCCCC system (except in cases on $n \le 6$ points, when the condition is vacuous). For it is easy to see that any hypertournament of rank 6 on $\{1, 2, \ldots, 7\}$ is obtained from the transitive hypertournament by negating points and possibly complementing; let's call this a pretransitive hypertournament. Say that the *weight* of a point is the number of smaller points, plus 1 if that point is negated. Fixing any point of a pretransitive hypertournament yields another pretransitive hypertournament in which the weights of all remaining points change parity. Therefore we can always fix three points so that the weights of the remaining four points are all even or all odd. Those four points violate Axiom 4. (For example, suppose the given sextuples are $123\bar{4}56$, $123\bar{4}57$, $123\bar{4}67$, $123\bar{5}67$, $12\bar{4}567$, $13\bar{4}567$, $23\bar{4}567$; this is the pretransitive hypertournament of rank 6 on $\{1, 2, 3, \bar{4}, \bar{5}, 6, 7\}$. The respective weights of 1, 2, 3, 4, 5, 6, 7 are 0, 1, 2, 4, 5, 5, 6, so we have four even weights and three odd weights. Fixing the three points of odd weight, namely $\{2, 5, 6\}$, will leave us with $2\bar{5}6\bar{1}3\bar{4} = 256143$, $2\bar{5}6\bar{1}37 = 256137$, $2\bar{5}6\bar{1}\bar{4}7 = 256174$, and $2\bar{5}6\bar{3}\bar{4}7 = 256347$; the triples 143, 137, 174, 347 violate Axiom 4. Fixing any other three points in this example would, however, produce a CC system.)

What about convex hulls in three dimensions? If we are given a set of points $p = (x_p, y_p, z_p)$, with no four coplanar, we can define a hypertournament by the rule

$$ pqrs \iff \det \begin{pmatrix} x_p & y_p & z_p & 1 \\ x_q & y_q & z_q & 1 \\ x_r & y_r & z_r & 1 \\ x_s & y_s & z_s & 1 \end{pmatrix} > 0\,. \tag{20.5} $$

Then the convex hull consists of all triangular faces Δpqr such that $spqr$ holds for all $s \notin \{p,q,r\}$. In the remainder of this section we shall let $|pqrs|$ denote the determinants in $|pqrs|$ (20.5), instead of considering the special case $z_p = x_p^2 + y_p^2$ that we used in (17.1) for the incircle test.

The quadruples $pqrs$ defined by (20.5) satisfy Axiom C5; hence they form a geometric hypertournament, and the triples obtained by fixing any point u form a pre-CC system. To verify this it is sufficient to consider the case $x_u = y_u = z_u = 0$, and to use the fact that we have

$$|tpq|\,|tsr| + |tqr|\,|tsp| + |trp|\,|tsq| = 0 \tag{20.6}$$

even when $|pqr|$ has the general form

$$\det \begin{pmatrix} x_p & y_p & z_p \\ x_q & y_q & z_q \\ x_r & y_r & z_r \end{pmatrix}. \tag{20.7}$$

(The left side of (20.6) is a polynomial, and we know that it is zero for all positive values of z_p, z_q, z_r, z_s, z_t; hence it must be identically zero.)

But the quadruples $pqrs$ of (20.5) do not form a CCC system in general, because they do not necessarily obey Axiom C4. Indeed, we obtain a violation of C4—a set of points such that

$$pqrt, \quad sprt, \quad sqpt, \quad \text{and} \quad sqrp \tag{20.8}$$

all are true—if and only if p is interior to the tetrahedron formed by $sqrt$, because

$$|pqrt| + |sprt| + |sqpt| + |sqrp| = |sqrt| \tag{20.9}$$

and Cramer's rule tells us that

$$p = \frac{|pqrt|}{|sqrt|}\,s + \frac{|sprt|}{|sqrt|}\,q + \frac{|sqpt|}{|sqrt|}\,r + \frac{|sqrp|}{|sqrt|}\,t. \tag{20.10}$$

(Identities (20.8) and (20.9) follow from the argument we used to derive (1.2) and (1.3). Points that make the determinant $|pqrs|$ positive form tetrahedrons that traditionally have a negative orientation, in classical treatments; our definition of $pqrs$ was chosen for consistency with the usual definitions of the counterclockwise and incircle predicates.)

The quadruples $pqrs$ do satisfy an axiom that is weaker than C4 but not deducible from C5:

Axiom C4′. $pqrt \wedge prst \wedge psqt \wedge srqt \implies pqrs$.

This law follows from (20.9) if we replace (p,q,r,s,t) respectively by (t,r,q,s,p). Let us say that a set of quadruples forms a *weak CCC system* if it satisfies C1, C2, C3, C4′, and C5.

It turns out that every weak CCC system has a 3D convex hull, i.e., a set of triangular faces Δpqr such that $spqr$ holds for all $s \notin \{p, q, r\}$. Notice that pq is an edge of such a triangle if and only if we have sr for all s in the vortex-free tournament obtained by fixing pq; this happens iff pq has a transitive tournament, with r the "champion" of that tournament. The tournament obtained by fixing pq is the complement of the tournament fixing qp; hence both are transitive or both are nontransitive. Therefore every edge pq of a 3D convex hull is part of exactly two triangles; every triangle Δpqr has three neighbors of the fours $\Delta qpr'$, $\Delta rqp'$, and $\Delta prq'$, just as we have observed in Delaunay triangulations. In fact, the main difference between the Delaunay triangulation of a CCC system and the 3D convex hull of a weak CCC system is the fact that all points p participate in the Delaunay triangulation, while the points p of the 3D hull are those whose associated pre-CC system is a genuine CC system.

Theorem. *Every weak CCC system on 3 or more points has a 3D convex hull, which is a set of triangles topologically equivalent to the surface of a sphere. A point p is part of the 3D hull if and only if we get a CC system when we fix p; this happens iff p does not satisfy (20.8) for any other points q, r, s, and t.*

Proof. We proceed by induction on the total number of points, n. If $n = 3$, there is trivially a convex hull consisting of two triangles Δpqr and Δprq; hence we assume that $n > 3$. Let p be one of the points, and suppose the 3D hull of the other $n - 1$ points has been found.

If the triples obtained by fixing p form a CC system, then they have a convex hull $t_1 t_2, \ldots, t_{m-1} t_m, t_m t_1$; hence the triangles $\Delta p t_2 t_1, \ldots, \Delta p t_m t_{m-1}, \Delta p t_1 t_n$ are part of the 3D convex hull. Each of these triangles $\Delta p t_j t_{j-1}$ is adjacent to a triangle $\Delta t_{j-1} t_j q_j$ of the 3D convex hull on the remaining $n - 1$ points. Since each edge of a 3D hull is part of exactly two triangles, the new triangles must replace all triangles previously interior to the polygon $t_m t_{m-1}, \ldots, t_1$, and we must have $qrsp$ true for every such triangle. Thus the new convex hull remains topologically equivalent to a sphere.

Suppose the pre-CC system obtained by fixing p do not form a CC system, and let $\mathcal{H}$ be the 3D hull of the other $n - 1$ points. We want to prove that $\mathcal{H}$ is also the 3D hull of all n points; in other words, if Δqrs is any triangle of $\mathcal{H}$ we want to prove that $pqrs$ holds. If not, we have $qrsp$, and we also have $rqtp$ where Δrqt is any triangle adjacent to Δqrs in $\mathcal{H}$. Since $\mathcal{H}$ is connected, we must in fact have $qrsp$ for all $\Delta qrs \in \mathcal{H}$. Therefore if we negate p, we obtain a geometric hypertournament H' in which $\mathcal{H}$ is the convex hull.

We will prove below that every geometric hypertournament of rank 4 having a nonempty 3D convex hull is a weak CCC system, i.e., satisfies Axiom C4'. This will lead to the desired contradiction. For we have assumed that the pre-CC system fixing p violates Axiom 4; hence there are points q, r, s, t satisfying (20.8). But Axiom C4 holds by assumption, hence we have

$$pqrt \wedge prst \wedge psqt \wedge psrq \wedge tsrq \,.$$

Therefore, negating p, we have

$$\bar{p}rqt \,\wedge\, \bar{p}srt \,\wedge\, \bar{p}qst \,\wedge\, \bar{p}qrs \,\wedge\, tsrq .$$

But this is a counterexample to C4′ in H'.

Therefore the proof of the theorem has been reduced to proving a sort of converse: Let H be any geometric hypertournament of rank 4 having a nonempty 3D convex hull. We will show that Axiom C4′ holds in H. Fix ab in H, where $\triangle abc$ is part of that hull, thereby obtaining a transitive tournament. The corollary in section 5 now tells us that the pre-CC systems obtained by fixing points a and b are in fact CC systems. Therefore if Axiom C4′ fails for some points p, q, r, s, and t, we have

$$pqrt \,\wedge\, prst \,\wedge\, psqt \,\wedge\, srqt \,\wedge\, spqr , \qquad (20.11)$$

and a or b cannot be in $\{p, q, r, s, t\}$.

Suppose t is the maximum element among $\{p, q, r, s, t\}$ in the transitive tournament for ab. Then we have $atbp$, $atbq$, $atbr$, and $atbs$; so if we consider the string of signed points $\alpha_1 \ldots \alpha_{n-2}$ that defines the vortex-free tournament for at, where $\alpha_1 = b$, we see that points p, q, r, s must occur in that string with a positive sign. This means edge at is part of the 3D convex hull if we restrict consideration to the six points $\{a, p, q, r, s, t\}$. But that cannot be, because the pre-CC system fixing t on those points includes either the triples $rqp \wedge srp \wedge qsp \wedge qrs$ or $pqr \wedge prs \wedge psq \wedge srq$. (We have used the fact that (20.11) has 60-fold symmetry: Interchanging any two points complements all the quadruples.) $\square$

Our proof shows, in fact, that it is possible to take any geometric hypertournament of rank 4 and negate a subset of points so as to obtain a weak CCC system. Let p, q, r be arbitrary points, and negate the other points if necessary so that the tournament for pq is transitive with maximum point r. Then $\triangle pqr$ is in the 3D hull, so Axiom C4′ must hold in the resulting system.

From these observations we can modify the dag triangulation algorithm of section 18 to obtain a general algorithm for 3D convex hulls in any weak CCC system. Whenever we find a point satisfying $qrsp$ for some $\triangle qrs$ in the current triangulation, we know that p will be part of the convex hull of the points seen so far. Whenever we find a point p not satisfying Axiom C4, therefore satisfying (20.8) for some q, r, s, and t, we know that p does not affect the convex hull so we can simply drop it. Complications arise when the starting '∞' is dropped; then we need to use a new point in its place. But in practice if we are using coordinates we can run the algorithm twice, once with $\infty = (0, 0, +\infty)$ and once with $\infty = (0, 0, -\infty)$; these infinite points will not be dropped, and we can piece together the overall convex hull from the two half-hulls if no two points have the same x and y coordinates. Further details are left to the reader.

To complete our study we will show that geometric hypertournaments of rank r are essentially equivalent to uniform oriented matroids of rank r, just as we showed in section 10 that pre-CC systems are in two-to-one correspondence with 4M systems.

Indeed, we need not repeat the proofs of section 10, because the same arguments carry through in general. It will suffice to indicate how the basic definitions are modified to cover geometric hypertournaments of arbitrary rank.

An $(r+1)$M system is like a 4M system, except that each circuit has $r+1$ signed points instead of 4. Our main connecting link between CC systems and 4-element circuits of oriented matroids was equation (10.1), which said that $\{p, q, r, s\}$ is a circuit if and only if

$$sqp = srq = spr = pqr.$$

The corresponding equation for 3-element circuits $\{p, q, r\}$ is simply

$$pq = qr = rp;$$

and for 5-element circuits $\{p, q, r, s, t\}$ it is

$$pqrs = qrst = rstp = stpq = tpqr.$$

In general for $(r+1)$M systems, $\{p_0, \ldots, p_r\}$ will be a circuit if and only if we have

$$p_1 \ldots p_r = \neg p_0 p_2 \ldots p_r = p_0 p_1 p_3 \ldots p_r = \cdots = (\neg)^r p_0 \ldots p_{r-1} \qquad (20.12)$$

in the corresponding geometric hypertournament. (These r-tuples are the respective duals of the rank 1 hypertournament on $\{p_0, \ldots, p_r\}$ with p_k true iff k is even.)

The operation of fixing a point p in an $(r+1)$M system consists simply of deleting all circuits that do not contain p or $\bar{p}$, then removing p or $\bar{p}$ from the remaining circuits. This is easily seen to correspond to the operation of fixing p in the associated hypertournament. To show that every $(r+1)$M system defines two complementary hypertournaments, we argue as in section 10 that the definition is unambiguous except for the truth value of a single r-tuple. Conversely, to prove that a geometric hypertournament of rank r yields an $(r+1)$M system, we prove as before that it yields an $(r+1)$L system. Axiom L3 holds because we know that every geometric hypertournament of rank r on $r+2$ elements is preisomorphic to the transitive hypertournament of rank r.

A CCC system is a 5M system (i.e., a uniform oriented matroid of rank 4) in which all circuits contain two elements of one sign and three elements of the other sign. A weak CCC system is an acyclic 5M system; this means that the circuits each contain at least one positive and one negative element. A CCCC system is a 6M system in which every circuit contains three elements of each sign. It is well known that acyclic $(r+1)$M systems correspond to convex sets in $r-1$ dimensions with vertices in "general position." The definitions given here, via hypertournaments satisfying the vortex-free property under projections, seem to be the simplest way to characterize the properties of those sets.

21. Historical remarks

The issues discussed above have been investigated by many authors, and the beautiful properties enjoyed by configurations of points have been expressed in a variety of interesting ways. The first paper on the subject, as far as we know today, was the remarkably prescient 19th-century work by R. Perrin [60], who described sequences of permutations that are equivalent to the notion of reflection networks. The next step was taken independently by Friedrich W. Levi in the 1920s [52], who analyzed the properties of what he called Pseudogerade (in German). An excellent summary of related work on arrangements of lines and pseudolines during the next 40 years appeared in books by Branko Grünbaum [34, 35], whose research and teaching stimulated considerable activity in the 70s and 80s.

Goodman and Pollack [26] resurrected Perrin's permutations and called them *allowable sequences*; they pointed out that nonoverlapping transpositions cannot always be interchanged without affecting realizability. Shortly afterwards [27, 28] they discovered that allowable sequences could be used to characterize arrangements of lines and pseudolines as well as counterclockwise orientations; and they generalized allowable sequences to account for degenerate cases, by permitting more than two adjacent elements to be flipped and by allowing several nonoverlapping flips to be performed simultaneously. They noted that the "betweenness rule" (7.1) was necessary [27, Proposition 2.16(c)], but they did not at first observe that it was also sufficient. In [30] they studied several other equivalent formulations of arrangements, notably *semispaces* (i.e., the sets of points that can occupy the top k lines of a reflection network, for some k), and what they called *generalized configurations of points* (which are essentially dual to pseudolines). Many of the results in sections 5–8 above are equivalent to theorems proved in [30]. In particular, transformations that preserve isomorphism, as discussed in section 8, are the subject of their Theorem 1.7; the fact that we can obtain all preisomorphic CC systems by a sequence of local transformations, as discussed at the end of section 5, is their Theorem 3.8. The lemma of section 9, which establishes the asymptotic formula $B_n = 2^{\Omega(n^2)}$, is essentially equivalent to a construction given in [29, Proposition 6.2]. A comprehensive survey of the theory of allowable sequences has just been published [32].

Axioms more closely related to the axioms for CC systems and pre-CC systems in section 1 were first proposed by L. Gutierrez Novoa in 1965 [39], who called his systems *n-ordered sets*. An $(r-1)$-ordered set is a mapping of r-tuples $p_1 \ldots p_r$ into $\{-1, 0, +1\}$ having two properties:

G1. Any interchange of two points $p_j \leftrightarrow p_k$ negates the value of $p_1 \ldots p_r$.

G2. If $q_1 p_2 \ldots p_r \cdot p_1 q_2 \ldots q_r \geq 0$ and $q_2 p_2 \ldots p_r \cdot q_1 p_1 q_3 \ldots q_r \geq 0$ and $\cdots$ and $q_r p_2 \ldots p_r \cdot q_1 \ldots q_{r-1} p_1 \geq 0$, then $p_1 p_2 \ldots p_r \cdot q_1 q_2 \ldots q_r \geq 0$.

For example, a 1-ordered set is a mapping from ordered pairs to $\{-1, 0, 1\}$ with $pq = -qp$, such that

$$qp' \cdot pq' \geq 0 \quad \wedge \quad q'p' \cdot qp \geq 0 \quad \Longrightarrow \quad pp' \cdot qq' \geq 0. \qquad (21.1)$$

In particular, if we have $pq \neq 0$ whenever $p \neq q$, this axiom is equivalent to saying that the elements form a vortex-free tournament. (To prove this, it suffices to consider a 4-point subset.) The general solution to (21.1) is slightly more complicated: If we say that $p \equiv q$ whenever we have either $px = qx$ for all x or $px = -qx$ for all x, then the equivalence classes form a vortex-free tournament, with the possible addition of one class having $px = 0$ for all x.

A 2-ordered set is a mapping from triples that satisfies

$$qp'p'' \cdot pq'q'' \geq 0 \quad \wedge \quad q'p'p'' \cdot qpq'' \geq 0 \quad \wedge \quad q''p'p'' \cdot qq'p \geq 0$$
$$\implies pp'p'' \cdot qq'q'' \geq 0. \qquad (21.2)$$

Setting $q'' = p''$ shows that we obtain a 1-ordered set if we fix the point p''; thus, if $pqr \neq 0$ for all distinct points $\{p, q, r\}$, we obtain the equivalent of Axiom 5. If we assume that there are points $p'p''$ such that $xp'p'' > 0$ for all $x \notin \{p', p''\}$—this says that the convex hull is nonempty—then (21.2) also yields Axiom 4. It can be shown that $(r - 1)$-ordered sets are equivalent to oriented matroids of rank r [48, 50].

Related ideas were developed independently about 15 years later by stereo-chemists at the University of Zurich, where André Dreiding proposed the name *chirotope* to describe arrangements of molecules in space; this name is suggestive because molecules without reflective symmetry have traditionally been called "chiral." The term *hypertournament* was also coined by members of the same group [71]. An early form of their definition was published on page 53 of [15], where the authors gave conditions approximately equivalent to Axioms C1, C2, C3, C4', and C5 in section 20 above, but the rules were stated in terms of the signs of determinant-like functions. A more refined definition of chirotope was subsequently published by Bokowski and Sturmfels [9]: A chirotope of rank r is a mapping from r-tuples $p_1 \ldots p_r$ to $\{-1, 0, 1\}$ satisfying Axiom G1 above and the following property in place of G2: If we fix the values of any $r - 2$ points $t_1, \ldots, t_{r-2}$, the resulting pair function $pq = t_1 \ldots t_{r-2}pq$ satisfies

$$\{sp \cdot qr, sq \cdot rp, sr \cdot pq\} = \{0\} \text{ or } \{-1, +1\} \text{ or } \{-1, 0, +1\} \qquad (21.3)$$

for all p, q, r, s. A *simplicial chirotope*, which has the additional property that $p_1 \ldots p_r$ is never 0 when the points p_j are distinct, is therefore precisely equivalent to what we have called a geometric hypertournament. In general, when $p_1 \ldots p_r$ can be 0, chirotopes define systems slightly more general than $(r-1)$-ordered sets; if a chirotope is not an $(r - 1)$-ordered set, we can remove some subset of its points and fix another subset so as to obtain k-tuples on a set of $2k$ elements $\{p_1, \ldots, p_k, q_1, \ldots, q_k\}$ for some $k \geq 3$, such that $p_1 \ldots p_k = q_1 \ldots q_k = 1$ and the value of all other k-tuples is zero. Such systems, discussed by Dress [14], violate G2 but satisfy (21.3).

Oriented matroids are based on the theory of ordinary matroids, introduced by Hassler Whitney in 1935 as an abstraction of the common notion of linear dependence [70]. Where Whitney retained only the zero/nonzero aspect of coefficients in vector equations, the oriented theory went a bit further and retained the sign of each coefficient. R. T. Rockafellar suggested in 1967 that such an approach might be fruitful [63], and oriented matroids were discovered a few years later, independently by Jon Folkman, Robert G. Bland, and Michel Las Vergnas. Folkman died tragically before being able to complete a paper on the subject; Bland and Las Vergnas learned of each other's work in 1975 and published a joint paper [6]. Meanwhile Jim Lawrence was preparing Folkman's notes for publication, and he extended them in several ways [21], notably by showing that oriented matroids of rank 3 correspond to arrangements of pseudolines and that oriented matroids of arbitrary rank correspond to arrangements of what he called "pseudo-hemispheres." Lawrence also introduced the notion of a *gatroid*, in which some elements are signed and others are signless.

Oriented matroids of rank r in which all r-element sets are independent have been called *free* [6], *simple* [21], or *uniform* [41]; of these three terms, "simple" might seem best, because it corresponds to simple arrangements and to simplicial chirotopes, but unfortunately it has already been used to describe oriented matroids whose circuits all have size 3 or more. Therefore "uniform" is currently the adjective of choice. Uniform oriented matroids, like the simplicial chirotopes, are completely characterized by the excluded-minor property that we have called vortex-freeness; in other words, they correspond to our Axiom 5. The other key ingredient of CC systems, Axiom 4, is equivalent to a special class of uniform oriented matroids that Las Vergnas [49] called *acyclic*, as defined in section 10 above. Las Vergnas showed that every oriented matroid can be made acyclic by suitable negation of points; we have seen special cases of this general result in sections 5 and 20. He developed a general theory of convexity for acyclic oriented matroids, and A. Mandel [53] extended this theory by proving (among other things) that the face lattice of any acyclic rank r oriented matroid, realizable or not, is isomorphic to the lattice of faces of a piecewise linear $(r-2)$-sphere. The theorems in sections 11 and 20 about the existence of convex hulls in CC systems and 3D convex hulls in weak CCC systems are simple special cases of this much more general theory.

Substantial research has been done on the question of deciding when an arrangement of pseudolines is "stretchable" to an isomorphic arrangement of straight lines; this is equivalent to deciding when a CC system is realizable by points in the Cartesian plane. Levi [52] observed that non-stretchable arrangements exist, if we allow three pseudolines to intersect at the same point; such arrangements correspond to nonrealizable oriented matroids of rank 3 in which some 3-element subsets are dependent. Ringel [62] exhibited 9 pseudolines in a nonstretchable *simple* arrangement, i.e., an arrangement in which all intersections between pairs of pseudolines are distinct. Ringel's example is equivalent to a nonrealizable CC system on 9 points. Goodman and Pollack [27] proved that all CC systems on 8 or fewer points are realizable; but they proved a few years later [31] that almost all CC systems on n points are unrealizable, in the limit as $n \to \infty$. Bokowski and Sturmfels [10] constructed nonrealizable CC systems on $10, 12, 14, \ldots$ points with the property that a realizable system is obtained when any point is deleted. Hence there can be no finite set of axioms analogous to Axioms 1–5 that characterize precisely the realizable systems.

The upper bound $2^{O(n^2)}$ on the total number of CC systems was known to researchers in computational geometry because of "horizon theorems" discovered independently by Chazelle, Guibas, and Lee [11, Lemma 1] and by Edelsbrunner, O'Rourke, and Seidel [18, Theorem 2.7]. The latter paper also claimed, in essence, that there are $2^{O(n^{r-1})}$ isomorphism classes of oriented matroids of rank r, for arbitrary r, and a proof of this claim was discovered in 1991 [19]. A recent paper of Bern, Eppstein, Plassman, and Yao [3] has established a sharper horizon theorem in the plane, namely that the total number of sides in all cells cut by an $(n+1)$st pseudoline is at most $9.5n - 1$. We can restate this in the terminology of section 9 above by considering the dag of all cutpaths defined in connection with the theorem of that section: The total of indegrees plus outdegrees of all vertices on any path from the source to the sink in that dag is at most $5.5n + m - 5$, where m is the number of extreme points (i.e., the outdegree of the source vertex, also the indegree of the sink vertex). Marshall Bern [4] has observed that the constant 3 in the theorem of section 9 can therefore be improved to $\sqrt[4]{54}$, which is approximately 2.711. On the other hand, as noted in that section, the best possible constant may well be 2, because no examples are yet known of reflection networks for which the total number of cutpaths exceeds $n2^{n-2}$.

The numbers E_n in section 9 have been widely studied, and calculated by methods independent of those in section 8. Halsey found $E_8 = 135$ in 1971 [40]; this value was confirmed by computer calculations carried out by Richter in 1987 (see [61]) and independently by Gonzales-Sprinberg and Laffaille in 1989 [25]. Richter and Laffaile continued the calculations to obtain $E_9 = 4382$, and Laffaile obtained also the value $E_{10} = 312356$. Richter-Gebert showed that all but 1 of the 4382 equivalence classes on 9 points are realizable; the unique unrealizable 9-point arrangement of pseudolines, is in fact the only one whose premutations are independent, in the sense discussed at the end of section 7 above. Bokowski and Richter-Gebert subsequently determined that precisely 242 of the classes on 10 points are unrealizable [7].

Hypertournaments of rank r are called "abstract binary $(r-1)$D-configurations" by Klin, Tratch, and Zefirov [44], who use Pólya's enumeration theorem to count the number of nonisomorphic hypertournaments of rank 3 on n vertices.

CCC systems are called *uniform matroid polytopes of rank 4* in the literature of oriented matroids; CCCC systems are called *neighborly matroid polytopes of rank 5*. Preisomorphism is called *reorientation equivalence*.

A definitive reference book covering almost all that is currently known about oriented matroids is scheduled to be published shortly [5].

22. Open problems

Several natural questions are suggested by the topics discussed above, and it may be worthwhile to list here some of the more interesting issues that the author has not had time to resolve:

1.   This monograph was originally motivated by the work of Guibas and Stolfi [38], who presented an algorithm for Delaunay triangulation written entirely in terms of the CC and InCircle predicates. After reading that paper, the author developed a craving for a firm understanding of those predicates, so that formal proofs of such algorithms could be given.

Further investigation revealed that the algorithm of [38] also has a hidden dependence on the coordinates of the points, because it relies on a procedure for dividing a set of points into two approximately equal halves. We have seen in section 14 that other concepts like lexicographic order can interact with the CC predicate in interesting ways; yet it remains interesting to ponder algorithms that do not use anything more than counterclockwise tests pqr to make all of their decisions. Therefore it is natural to wonder if there is an $O(N)$ algorithm to partition an arbitrary CC system into approximately equal semispaces.

Related questions can also be posed: Is it possible to find an extreme point of an arbitrary CC system in $O(N)$ time? How long does it take to test whether or not a given pre-CC system is a CC system?

2. The treehull algorithm in section 11 finds the convex hull of any given CC system in $O(N \log N)$ steps. Is there an analogous incremental algorithm that will find the Delaunay triangulation of any given CCC system, with worst-case time $O(N \log N)$? Of particular interest would be a data structure that allows "amortized binary search" in an evolving triangulation.

3. The dag triangulation algorithm of section 18 runs in time $O(N \log N)$, on the average, but the upper bound derived in (18.12) is much larger than observed in practice. Is it possible to prove sharper estimates, more like those in section 12?

4. The dag triangulation algorithm is not parsimonious, in the sense of section 15. For example, it sometimes tests twice whether the new point p falls in the scope of a certain triangle, if that triangle is adjacent to two of the edges of the polygon surrounding p. Is there an efficient, parsimonious algorithm for Delaunay triangulation in a given CCC system?

5. What is the asymptotic number of CCC systems and CCCC systems definable on n points, as $n \to \infty$? (Grünbaum [34, §7.2.4] constructed three nonisomorphic CCCC systems on 8 points; Bland and Las Vergnas [6, Proposition 3.9] demonstrated the existence of "alternating orientations" that are not transitive. Shemer [65] showed that the number of CCCC systems realizable in Euclidean 4-space is $2^{\Omega(n \log n)}$, and he also observed that the total number of CCCC systems on n points is $2^{O(n^2)}$.)

6. Is every CCC system embeddable in a CCCC system?

7. Is there a simple relation between the Delaunay triangulation of a CCC system, as defined in section 17, and the "$\mathcal{D}$-Delaunay triangulations of a family $\mathcal{D}$ of pseudo-disks," as defined in [54]?

8. Is there a system of axioms on r-ary relations that can be satisfied over n points in a total of, say, $2^{\Theta(n \log n \log \log n)}$ ways? What orders of growth are possible in asymptotic enumeration formulas analogous to (4.11) and to the corollary that follows (9.7)?

9. What is a good algorithm to find the generalized convex hull of an $(r+1)$M system, i.e., a uniform acyclic oriented matroid of rank r? (When $r = 3$ this is the convex hull of a CC system, so we know the answer. When $r = 4$ it is the convex hull of a weak CCC system, generalizing the 3D convex hull; the algorithm sketched after the theorem in section 20 is incomplete.)

Acknowledgments

MANY PEOPLE provided considerable help to the author as these notes were being prepared, notably Eli Goodman, Leo Guibas, Ricky Pollack, Jürgen Richter-Gebert, David Salesin, Raimund Seidel, Bernd Sturmfels, Frances Yao, and Günter Ziegler. Special thanks are also due to Phyllis Winkler, who transformed more than 150 pages of scribbled manuscript into a respectable-looking scientific document. Some of the research was done during a visit to the Institute of Systems Science at the University of Singapore; the work was completed during a visit to Institut Mittag-Leffler in Djursholm, Sweden.

Bibliography

[1]   E. al-Aamily, A. O. Morris, and M. H. Peel, "The representations of the Weyl groups of type B_n," *Journal of Algebra* **68** (1981), 298–305. Cited on page 17.

[2]   Cecilia R. Aragon and Raimund G. Seidel, "Randomized search trees" (extended abstract), *30th IEEE Symposium on Foundations of Computer Science* (1989), 540–546. Cited on page 53.

[3]   Marshall Bern, David Eppstein, Paul Plassman, and Frances Yao, "Horizon theorems for lines and polygons," in *Discrete and Computational Geometry: Papers from the DIMACS Special Year*, edited by Jacob E. Goodman, Richard Pollack, and William Steiger, *DIMACS Series in Discrete Mathematics and Theoretical Computer Science* **6** (1991), 45–66. Cited on page 96.

[4]   Marshall Bern, personal communication, January 1991. Cited on page 97.

[5]   Anders Björner, Michel Las Vergnas, Bernd Sturmfels, Neil White, and Günter M. Ziegler, *Oriented Matroids*, Encyclopedia of Mathematics Series, Cambridge University Press (1992). Cited on page 97.

[6]   Robert G. Bland and Michel Las Vergnas, "Orientability of matroids," *Journal of Combinatorial Theory* **B24** (1978), 94–123. Cited on pages 40, 95, 96, and 98.

[7]   J. Bokowski, G. Laffaille, and J. Richter-Gebert, "10 point oriented matroids and projective incidence theorems," in preparation. Cited on page 97.

[8]   Jürgen Bokowski, Jürgen Richter, and Bernd Sturmfels, "Nonrealizability proofs in computational geometry," *Discrete & Computational Geometry* **5** (1990), 333–350. Cited on page 6.

[9]   Jürgen Bokowski and Bernd Sturmfels, "On the coordinatization of oriented matroids," *Discrete & Computational Geometry* **1** (1986), 293–306. Cited on page 95.

[10] Jürgen Bokowski and Bernd Sturmfels, "An infinite family of minor-minimal nonrealizable 3-chirotopes," *Mathematische Zeitschrift* **200** (1989), 583–589. Cited on page 96.

[11] Bernard Chazelle, Leonidas J. Guibas, and D. T. Lee, "The power of geometric duality," *BIT* **25** (1985), 76–90. Cited on page 96.

[12] Kenneth L. Clarkson and Peter W. Shor, "Applications of random sampling in computational geometry, II," *Discrete & Computational Geometry* **4** (1989), 387–421. Cited on page 81.

[13] B. Delaunay, "Neue Darstellung der geometrischen Krystallographie," *Zeitschrift für Kristallographie* **84** (1932), 109–149; errata, **85** (1933), 332. Cited on page 69.

[14] Andreas Dress, "Chirotops and oriented matroids: Diskrete Strukturen, algebraische Methoden und Anwendungen," *Bayreuther Mathematische Schriften* **21** (1986), 14–68. Cited on page 95.

[15]   Andreas Dress, André Dreiding, and Hans Haegi, "Classification of mobile molecules by category theory," in *Symmetries and Properties of Non-Rigid Molecules*, Proceedings of an International Symposium in Paris, France, 1–7 July 1982, edited by J. Maruani and J. Serre; *Studies in Physical and Theoretical Chemistry* **23** (1983), 39–58. Cited on page 95.

[16]   P. H. Edelman and C. Greene, "Balanced tableaux," *Advances in Mathematics* **63** (1987), 42–99. Cited on page 35.

[17]   Herbert Edelsbrunner and Ernst Peter Mücke, "Simulation of Simplicity: A technique to cope with degenerate cases in geometric algorithms," *Fourth Annual ACM Symposium on Computational Geometry* (1988), 118–133. Cited on page 59.

[18]   H. Edelsbrunner, J. O'Rourke, and R. Seidel, "Constructing arrangements of lines and hyperplanes with applications," *SIAM Journal on Computing* **15** (1986), 341–363. Cited on page 96.

[19]   H. Edelsbrunner, R. Seidel, and M. Sharir, "On the zone theorem for hyperplane arrangements," *SIAM Journal of Computing*, to appear. Preprint in *New Results and New Trends in Computer Science*, edited by Hermann Maurer, *Lecture Notes in Computer Science* **555** (1991), 108–123. Cited on page 96.

[20]   Robert W Floyd, personal communication, February 1964. Cited on page 29.

[21]   Jon Folkman and Jim Lawrence, "Oriented matroids," *Journal of Combinatorial Theory* **B25** (1978), 199–236. Cited on pages 40, 43, and 96.

[22]   Steven Fortune, "Stable maintenance of point set triangulations in two dimensions," *30th IEEE Symposium on Foundations of Computer Science* (1989), 494–499. Cited on pages 62 and 67.

[23]   Fred Galvin, personal communications, November 1991 and January 1992. Cited on page 15.

[24]   Michael R. Garey and David S. Johnson, *Computers and Intractability* (San Francisco: W. H. Freeman, 1979). Cited on page 20.

[25]   Gérard Gonzales-Sprinberg and Guy Laffaille, "Sur les arrangements simples de huit droites dans RP^2," *Comptes Rendus de l'Académie des Sciences*, Série I, **309** (1989), 341–344. Cited on page 97.

[26]   Jacob E. Goodman and Richard Pollack, "On the combinatorial classification of nondegenerate configurations in the plane," *Journal of Combinatorial Theory* **A29** (1980), 220–235. Cited on page 94.

[27]   Jacob E. Goodman and Richard Pollack, "Proof of Grünbaum's conjecture on the stretchability of certain arrangements of pseudolines," *Journal of Combinatorial Theory* **A29** (1980), 385–390. Cited on pages 94 and 96.

[28]   Jacob E. Goodman and Richard Pollack, "A theorem of ordered duality," *Geometriæ Dedicata* **12** (1982), 63–74. Cited on page 94.

[29]   Jacob E. Goodman and Richard Pollack, "Multidimensional sorting," *SIAM Journal on Computing* **12** (1983), 484–507. Cited on pages 46 and 94.

[30]   Jacob E. Goodman and Richard Pollack, "Semispaces of configurations, cell complexes of arrangements," *Journal of Combinatorial Theory* **A37** (1984), 257–293. Cited on pages 35 and 94.

[31]   Jacob E. Goodman and Richard Pollack, "Upper bounds for configurations and polytopes in R^d," *Discrete & Computational Geometry* **1** (1986), 219–227. Cited on pages 40 and 96.

[32]   Jacob E. Goodman and Richard Pollack, "Allowable sequences and order types in discrete and computational geometry," *New Trends in Discrete and Computational Geometry*, edited by J. Pach (Springer-Verlag, 1992), to appear. Cited on page 94.

[33]   Ronald L. Graham, Donald E. Knuth, Oren Patashnik, *Concrete Mathematics* (Reading, Mass.: Addison–Wesley, 1989). Cited on page 14.

[34]   Branko Grünbaum, *Convex Polytopes* (London Interscience, 1967). Cited on pages 94 and 98.

[35]   Branko Grünbaum, *Arrangements and Spreads.* Conference Board of the Mathematical Sciences, Regional Conference Series in Mathematics, Volume 10 (Providence, RI: American Mathematical Society, 1972). Cited on pages 34 and 94.

[36]   Leonidas J. Guibas, Donald E. Knuth, and Micha Sharir, "Randomized incremental construction of Delaunay and Voronoi diagrams," *Algorithmica* **7** (1992), 381–413. Abbreviated version in *Automata, Languages and Programming*, edited by M. S. Paterson, *Lecture Notes in Computer Science* **443** (1990), 414–431. Cited on pages 2, 3, 74, 77, and 80.

[37]   Leonidas Guibas, David Salesin, and Jorge Stolfi, "Constructing strongly convex approximate hulls with inaccurate primitives," *Algorithmica*, to appear. Abbreviated version in *Proceedings of the International Symposium on Algorithms SIGAL 90*, edited by T. Asano, T. Ibaraki, H. Imai, and T. Nishizeki, *Lecture Notes in Computer Science* **450** (1990), 261–270. Cited on page 67.

[38]   Leonidas Guibas and Jorge Stolfi, "Primitives for the manipulation of general subdivisions and the computation of Voronoi diagrams," *ACM Transactions on Graphics* **4** (1985), 74–123. Cited on pages v, 69, 72, and 97.

[39]   Lino Gutierrez Novoa, "On n-ordered sets and order completeness," *Pacific Journal of Mathematics* **15** (1965), 1337–1345. Cited on page 94.

[40]   Eric Richard Halsey, *Zonotopal complexes on the d-cube*, Ph.D. dissertation, University of Washington, Seattle, WA (1972). Cited on page 97.

[41]   Beat Jaggi, Peter Mani-Levitska, Bernd Sturmfels, and Neil White, "Uniform oriented matroids without the isotopy property," *Discrete & Computational Geometry* **4** (1989), 97–100. Cited on page 96.

[42]   J. W. Jaromczyk and G. W. Wasilkowski, "Numerical stability of a convex hull algorithm for simple polygons," University of Kentucky technical report 177–90 (1990), 18 pp. Cited on page 67.

[43] Arne Jonassen and Donald E. Knuth, "A trivial algorithm whose analysis isn't," *Journal of Computer and System Sciences* **16** (1978), 301–322. Cited on page 55.

[44] Mikhail H. Klin, Serge S. Tratch, and Nikolai S. Zefirov, "2D-configurations and clique-cyclic orientations of the graphs $L(K_p)$," *Reports in Molecular Theory* **1** (1990), 149–163. Cited on page 97.

[45] Donald E. Knuth, *The Art of Computer Programming*, Volume 3: *Sorting and Searching* (Reading, MA: Addison-Wesley, 1973). Cited on pages 29 and 47.

[46] Donald E. Knuth, "Two notes on notation," *American Mathematical Monthly* **99** (1992), 403–422. Cited on page 14.

[47] Donald E. Knuth, *The Stanford GraphBase*, book in preparation. Cited on page 53.

[48] Michel Las Vergnas, "Bases in oriented matroids," *Journal of Combinatorial Theory* **B25** (1978), 283–289. Cited on pages 3, 40, and 95.

[49] Michel Las Vergnas, "Convexity in oriented matroids," *Journal of Combinatorial Theory* **B29** (1980), 231–243. Cited on page 96.

[50] Alain Lascoux and Marcel-Paul Schützenberger, "Structure de Hopf de l'anneau de cohomologie et de l'anneau de Grothendieck d'une variété de drapeaux," *Comptes Rendus des séances de l'Académie des Sciences*, Série I, **295** (1982), 629–633. Cited on page 35.

[51] Jim Lawrence, "Oriented matroids and multiply ordered sets," *Linear Algebra and Its Applications* **48** (1982), 1–12. Cited on pages 3 and 95.

[52] F. Levi, "Die Teilung der projektiven Ebene durch Gerade oder Pseudogerade," Berichte über die Verhandlungen der sächsischen Akademie der Wissenschaften, Leipzig, Mathematisch-physische Klasse **78** (1926), 256–267. Cited on pages 34, 94, and 96.

[53] Arnaldo Mandel, *Topology of Oriented Matroids*. Ph.D. thesis, University of Waterloo, Ontario, 1982. Cited on page 96.

[54] Jiří Matoušek, Raimund Seidel, and Emo Welzl, "How to net a lot with little: Small ϵ-nets for disks and halfspaces," *Discrete & Computational Geometry*, to appear. Preprint B90–04, Freie Universität Berlin, Fachbereich Mathematik, August 1990. Cited on page 98.

[55] Victor J. Milenkovic and Zhenyu Li, "Constructing strongly convex hulls using exact or rounded arithmetic," *Sixth Annual ACM Symposium on Computational Geometry* (1990), 235–243. Cited on page 67.

[56] John W. Moon, *Topics on Tournaments* (New York: Holt, Rinehart and Winston, 1968). Cited on pages 7 and 15.

[57] J. W. Moon, "Tournaments whose subtournaments are irreducible or transitive," *Canadian Mathematical Bulletin* **21** (1979), 75–79. Cited on page 15.

[58] Ernest Morris, *The History and Art of Change Ringing* (London: Chapman & Hall, 1931). Cited on page 29.

[59]   W. Nowacki, "Der Begriff 'Voronoischer Bereich'," *Zeitschrift für Kristallographie* **85** (1933), 331-332. Cited on page 69.

[60]   R. Perrin, "Sur le problème des aspects," *Bulletin de la Société Mathématique de France* **10** (1882), 103–127. Cited on page 94.

[61]   J. Richter, "Kombinatorische realisierbarkeitskriterien für orientierte Matroide," *Mitteilungen aus dem Mathematischen Seminar Giessen* **194** (1989), 113 pp. Cited on page 97.

[62]   G. Ringel, "Über Geraden in allgemeiner Lage," *Elemente der Mathematik* **12** (1957), 75–82. Cited on page 96.

[63]   R. T. Rockafellar, "The elementary vectors of a subspace of R^n," in *Combinatorial Mathematics and Its Applications*, edited by R. C. Bose and T. A. Dowling, Proceedings of a conference in Chapel Hill, North Carolina, April 10–14, 1967 (University of North Carolina Press, 1969), 104–127. Cited on page 95.

[64]   T. J. Schaefer, "The complexity of satisfiability problems," *Tenth Annual ACM Symposium on Theory of Computing* (1978), 216–226. Cited on page 20.

[65]   Ido Shemer, "Neighborly polytopes," *Israel Journal of Mathematics* **43** (1982), 291–314. Cited on page 98.

[66]   Daniel Dominique Sleator and Robert Endre Tarjan, "Self-adjusting binary search trees," *Journal of the ACM* **32** (1985), 652–686. Cited on page 53.

[67]   Richard P. Stanley, "On the number of reduced decompositions of elements of Coxeter groups," *European Journal of Combinatorics* **5** (1984), 359–372. Cited on page 35.

[68]   Alfred Tarski, *A Decision Method for Elementary Algebra and Geometry*, second revised edition (Berkeley and Los Angeles: University of California Press, 1951). Cited on page 23.

[69]   Georges Voronoï, "Nouvelles applications des paramètres continus à la théorie des formes quadratiques," *Journal für die reine und angewandte Mathematik* **133** (1907), 97–178; **134** (1908), 198–287; **136** (1909), 67–181. Cited on page 69.

[70]   Hassler Whitney, "On the abstract properties of linear dependence," *American Journal of Mathematics* **57** (1935), 509–533. Cited on page 95.

[71]   Karl Wirth, *Endliche Hyperturniere*. Dissertation, Eidgenossische Technische Hochschule, Zürich, 1978. Cited on page 95.

[72]   Günter M. Ziegler, personal communication, December 1991. Cited on page 72.

Index

[Several special notations needed in this volume are indexed under 'notation'.]

Lecture Notes in Computer Science

For information about Vols. 1–529
please contact your bookseller or Springer-Verlag